ABOUT THE BOOK

A geographic information system, geographical information system, or geospatial information system is any system that captures, stores, analyses, manages, and presents data that are linked to location. In the simplest terms, GIS is the merging of cartography, statistical analysis, and database technology. GIS may be used in archaeology, geography, cartography, remote sensing, land surveying, public utility management, natural resource management, precision agriculture, photogrammetry, urban planning, emergency management, landscape architecture, navigation, aerial video, and localized search engines. Remote sensing images are normally in the form of digital images. In order to extract useful information from the images, image processing techniques may be employed to enhance the image to help visual interpretation, and to correct or restore the image if the image has been subjected to geometric distortion, blurring or degradation by other factors. Agricultural resources data are collected y aircraft and satellite-mounted instruments, which receive reflected energy from target in some frequency of the electromagnetic spectrum. The process involved in electromagnetic remote sensing system namely, data acquisition and data analysis are outlined below and a schematic diagram of electromagnetic remote sensing process. This book should be useful to remote sensing scientists and engineers, geographers, geologists, ecologists and environmental scientists, agricultural and soil scientists and urban and regional planners. Regional, state and university remote sensing centres may like to have this book as well.

ABOUT THE AUTHOR

Rita Mishra, M.Sc. Agriculture and pursuing her Ph.D. She is presently working in Nalanda College, Bihar as Professor, she is an erudite scholar and seasoned teacher to the core is highly qualified and teaching the subject at a reputed civil service Academy based in Delhi for last seven years .She has attended a number of seminars and conferences and also organised several workshops for civil service students and published several articles in various reputed journals pertaining to problems of farmers, future of agriculture etc. She published more than 13 books on Agriculture fields.

GIS and Remote Sensing Analysis for Cropping Systems

Rita Mishra

AMIGA PRESS. INC
Delhi -110051

AMIGA PRESS INC.

H.O :- 28-A, Indra Park,

Chander Nagar, Delhi :- 110051 (India)

Tel :- 011-22504028, +91-98736-57000

Email :- amigapress@gmail.com
amigaeditorial@gmail.com
amigapub@gmail.com

Website :- www.amigapress.co.in

This edition is published by Amiga Press Inc.

ISBN : 978-93-87295-37-7

First Published 2019

Printed and bound by Replika Press Pvt. Ltd.

Preface

A geographic information system, geographical information system, or geospatial information system is any system that captures, stores, analyses, manages, and presents data that are linked to location. In the simplest terms, GIS is the merging of cartography, statistical analysis, and database technology. GIS may be used in archaeology, geography, cartography, remote sensing, land surveying, public utility management, natural resource management, precision agriculture, photogrammetry, urban planning, emergency management, landscape architecture, navigation, aerial video, and localized search engines.

As GIS can be thought of as a system, it digitally creates and "manipulates" spatial areas that may be jurisdictional, purpose or application oriented for which a specific GIS is developed. Hence, a GIS developed for an application, jurisdiction, enterprise, or purpose may not be necessarily interoperable or compatible with a GIS that has been developed for some other application, jurisdiction, enterprise, or purpose.

In Optical Remote Sensing, optical sensors detect solar radiation reflected or scattered from the earth, forming images resembling photographs taken by a camera high up in space. The wavelength region usually extends from the visible and near infrared (commonly abbreviated as VNIR) to the short-wave infrared (SWIR). Different materials such as water, soil, vegetation, buildings and roads reflect visible and infrared light in different ways. They have different colours and brightness when seen under the sun. The interpretation of optical images require the knowledge of the spectral reflectance signatures of the various materials (natural or man-made) covering the surface of the earth. There are also infrared sensors measuring the thermal infrared radiation emitted from the earth, from which the land or sea surface temperature can be derived.

Remote sensing is nothing but a means to get the reliable information about an object without being in physical contact with the object. It is on the observation of an object by a device separated from it by some distance utilizing the characteristics response of different objects to emissions in the electromagnetic energy is measured in a number of spectral bands for the purpose of identification of the object.

In such study single tabular form of data or map data is not sufficient enough which can provide can be, combined with information's obtained from existing maps and tabular data.

An alternative to grid sampling is targeted or zone sampling. The soil samples are located in homogeneous management zones instead of uniformly spaced grids. The zones are laid out using a process similar to computer based unsupervised image classification. Images obtained from multispectral remote sensors are taken of the vegetated areas of the field. The pixel digital numbers for each band are separated into statistically separable clusters that are classified into homogeneous zones. This cuts down on the soil, terrain, plant growth, and other variability within each area to be managed; thus fewer soil samples are needed for each area.

Agriculture resources are among the most important renewable, dynamic natural resources. Comprehensive, reliable and timely information on agricultural resources is very much necessary for a country like India whose mainstay of the economy is agriculture. Agriculture survey are presently conducted throughtout the nation in order to gather information and associated statistics on crops, rangeland, livestock and other related agricultural resources. These information of data are most importance for the implementation of effective management decisions at local, panchayat and district levels. In fact, agricultural survey is a backbone of planning and allocation of the limited resources to different sectors of the economy. With increasing population pressure throughout the nation and the concomitant need for increased agricultural production (food and fiber crops as well as livestock) there is a definite need for improved management of the nation agricultural resources. In order to accomplish this, it is first necessary to obtain reliable data on not only the types, but also the quality, quantity and location of these resources. The remote sensing techniques has been and it will continue to, a very important factor in the improvement of the present systems of acquiring and generating agricultural data.

This is a reference book. All the matter is just compiled and edited in nature, taken from the various sources which are in public domain.

This book should be useful to remote sensing scientists and engineers, geographers, geologists, ecologists and environmental scientists, agricultural and soil scientists and urban and regional planners. Regional, state and university remote sensing centres may like to have this book as well.

—*Editor*

Contents

1

An Introduction to Geographic Information System

INTRODUCTION

A geographic information system, geographical information system, or geospatial information system is any system that captures, stores, analyses, manages, and presents data that are linked to location. In the simplest terms, GIS is the merging of cartography, statistical analysis, and database technology. GIS may be used in archaeology, geography, cartography, remote sensing, land surveying, public utility management, natural resource management, precision agriculture, photogrammetry, urban planning, emergency management, landscape architecture, navigation, aerial video, and localized search engines.

As GIS can be thought of as a system, it digitally creates and "manipulates" spatial areas that may be jurisdictional, purpose or application oriented for which a specific GIS is developed. Hence, a GIS developed for an application, jurisdiction, enterprise, or purpose may not be necessarily interoperable or compatible with a GIS that has been developed for some other application, jurisdiction, enterprise, or purpose.

What goes beyond a GIS is a spatial data infrastructure, a concept that has no such restrictive boundaries. Therefore, in a general sense, the term describes any information system that integrates, stores, edits, analyses, shares, and displays geographic information for informing decision making. GIS applications are tools that allow users to create interactive queries, analyse spatial information, edit data, maps, and present the results of all these operations. Geographic information science is the science underlying the geographic concepts, applications and systems.

Applications

GIS technology can be used for: earth surface based scientific investigations; resource management, reference, and projections of a geospatial nature—both manmade and natural; asset management and location planning; archaeology; environmental impact study; infrastructure assessment and development; urban planning; cartography, for a thematic and/or time based purpose; criminology; GIS data development geographic history; marketing; logistics; population and demographic studies; prospectivity mapping; location attributes applied statistical analysis; warfare assessments; and other purposes. Examples of use are: GIS may allow emergency planners to easily calculate emergency response times and the movement of response resources in the case of a natural disaster; GIS might be used to find wetlands that need protection strategies regarding pollution; or GIS can be used by a company to site a new business location to take advantage of GIS data identified trends to respond to a previously under-served market. Most city and transportation systems planning offices have GIS parts.

GIS THROUGH HISTORY

Some 35,000 years ago, Cro-Magnon hunters drew pictures of the animals they hunted on the walls of caves near Lascaux, France. Associated with the animal drawings are track lines and tallies thought to depict migration routes. These early records followed the two-element structure of modern geographic information systems (GIS): a graphic file linked to an attribute database.

Today, biologists use collar transmitters and satellite receivers to track the migration routes of caribou and polar bears to help design programs to protect the animals. In a GIS, the migration routes were indicated by different colours for each month for 21 months. Researchers then used the GIS to superimpose the migration routes on maps of oil development plans to determine the potential for interference with the animals.

Mapmaking

Researchers are working to incorporate the mapmaking processes of traditional cartographers into GIS technology for the automated production of maps. One of the most common products of a GIS is a map. Maps are generally easy to make using a GIS and they are often the most effective means of communicating the results of the GIS process. Therefore, the GIS is usually a prolific producer of maps. The users of a GIS must be concerned with the quality of the maps produced because the GIS normally does not regulate common cartographic principles. One of these principles is the concept of generalization, which deals with the content and detail of information at various scales. The GIS user can change scale at the push of a button, but controlling content and detail is often not so easy. Mapmakers have long recognized that content and detail need to change as the scale of the map changes. For example, the State of New Jersey can be mapped at various scales, from the small scale of 1:500,000 to the larger scale of 1:250,000 and the yet larger scale of 1:100,000,but each scale requires an appropriate level of generalization.

Site Selection

The U.S. Geological Survey (USGS), in a cooperative project with the Connecticut Department of Natural Resources, digitized more than 40 map layers for the areas covered by the USGS Broad Brook and Ellington 7.5-minute topographic quadrangle maps. This information can be combined and manipulated in a GIS to address planning and natural resource issues. GIS information was used to locate a potential site for a new water well within half a mile of the Somers Water Company service area. To prepare the analysis, cartographers stored digital maps of the water service areas in the GIS. They used the proximity function in the GIS to draw a half-mile buffer zone around the water company service area. This buffer zone was the "window" used to view and combine the various map coverages relevant to the well site selection.

The land use and land cover map for the two areas shows that the area is partly developed. A GIS was used to select undeveloped areas from the land use and land cover map as the first step in finding well sites. The developed areas were eliminated from further consideration. The quality of water in Connecticut streams is closely monitored. Some of the streams in the study area were known to be unusable as drinking water sources. To avoid pulling water from these streams into the wells, 100-meter buffer zones were created around the unsuitable streams using the GIS, and the zones were plotted on the map. The areas in blue have the characteristics desired for a water well site.

Point sources of pollution are recorded by the Connecticut Department of Natural Resources. These records consist of a location and a text description of the pollutant. To avoid these toxic areas, a buffer zone of 500 meters was established around each point. This information was combined with the previous two map layers to produce a new map of areas suitable for well sites. Points sources of pollution in the water service area are identified and entered into a GIS.

The map of surficial geology shows the earth materials that lie above bedrock. Since the area under consideration in Connecticut is covered by glacial deposits, the surface consists largely of sand and gravel, with some glacial till and fine-grained sediments. Of these materials, sand and gravel are the most likely to store water that could be tapped with wells. Areas underlain by sand and gravel were selected from the surficial geology map. They were combined with the results of the previous selections to produce a map consisting of: (1) sites in underdeveloped areas underlain by sand and gravel, (2) more than 500 meters from point sources of pollution, and (3) more than 100 meters from unsuitable streams. A map that shows the thickness of saturated sediments was created by using the GIS to subtract the bedrock elevation from the surface elevation. For this analysis, areas having more than 40 feet of saturated sediments were selected and combined with the previous overlays.

The resulting site selection map shows areas that are undeveloped, are situated outside the buffered pollution areas, and are underlain by 40 feet or more of water-saturated sand and gravel. Because of map resolution and the limits of precision in digitizing, the very small polygons (areas) may not have all of the characteristics analysed, so another GIS function was used to screen out areas smaller than 10 acres. The final six sites are displayed with the road and stream network and selected place names for use in the field. Potential water well sites, roads, streams and place names. The process illustrated by this site selection analysis has been used for many common applications, including transportation planning and waste disposal site location. The technique is particularly useful when several physical factors must be considered and integrated over a large area.

Emergency Response Planning

A GIS was used to combine road network and earth science information to analyse the effect of an earthquake on the response time of fire and rescue squads. The area covered by the USGS Sugar House 7.5-minute topographic quadrangle map was selected for the study because it includes both undeveloped areas in the mountains and a part of Salt Lake City. Detailed earth science information was available for the entire region. The road network from a USGS digital line graph includes information on the types of roads, which range from rough trails to divided highways. The locations of fire stations were plotted on the road network. A GIS function called network analysis was used to calculate the time necessary for emergency vehicles to travel from the fire stations to different areas of the city. The network analysis function considers two elements: (1) distance from the fire station, and (2) speed of travel based on the type of road. The analysis shows that under normal conditions, most of the area within the city will be served in less than 7 minutes and 30 seconds because of the distribution and density of fire stations and the continuous network of roads.

The accompanying illustration depicts the blockage of the road network that would result from an earthquake, assuming that any road crossing the fault trace would become impassable. The primary effect on emergency response time would occur in neighbourhoods west of the fault trace, where travel times from the fire stations would be noticeably lengthened. Road network of area covered by the Sugar House quadrangle plotted from USGS digital line graph data, indicating the locations of fire stations and travel times of emergency vehicles. Areas in blue can receive service within 2½minutes, area in green within 5 minutes, areas in yellow within 7½ minutes, and areas in magenta within 10 minutes. Areas in white cannot receive service within 10 minutes.

After faulting, initial model. Network analysis in a GIS produces a map of travel times from the stations after faulting. The fault is in red. Emergency response times have increased for areas west of the fault. The Salt Lake City area lies on lake sediments of varying thicknesses.

These sediments range from clay to sand and gravel, and most are water-saturated. In an earthquake, these materials may momentarily lose their ability to support surface structures, including roads.

The potential for this phenomenon, known as liquefaction, is shown in a composite map portraying the inferred relative stability of the land surface during an earthquake. Areas near the fault and underlain by thick, loosely consolidated, water-saturated sediments will suffer the most intense surface motion during an earthquake. Areas on the mountain front with thin surface sediments will experience less additional ground acceleration. The map of liquefaction potential was combined with the road network analysis to show the additional effect of liquefaction on response times.

The final map shows that areas near the fault, as well as those underlain by thick, water-saturated sediments, are subject to more road disruptions and slower emergency response than are other areas of the city. Map of potential ground l liquefaction during an earthquake. The least stable areas are shown by yellows and oranges, the most stable by grays and browns. After faulting, final model. A map showing the effect of an earthquake on emergency travel times is reduced by combining the liquefaction potential information.

THREE-DIMENSIONAL GIS

To more realistically analyse the effect of the Earth's terrain, we use three-dimensional models within a GIS. A GIS can display the Earth in realistic, three-dimensional perspective views and animations that convey information more effectively and to wider audiences than traditional, two-dimensional, static maps.

The U.S. Forest Service was offered a land swap by a mining company seeking development rights to a mineral deposit in Arizona's Prescott National Forest. Using a GIS, the USGS and the U.S. Forest Service created perspective views of the area to depict the terrain as it would appear after mining. To assess the potential hazard of landslides both on land and underwater, the USGS generated a three-dimensional image of the San Francisco Bay area. It created the image by mosaicking eight scenes of natural colour composite Landsat 7 enhanced thematic mapper imagery on California fault data using approximately 700 digital elevation models at 1:24,000 scale.

Graphic Display Techniques

Traditional maps are abstractions of the real world; each map is a sampling of important elements portrayed on a sheet of paper with symbols to represent physical objects. People who use maps must interpret these symbols. Topographic maps show the shape of the land surface with contour lines. Graphic display techniques in GISs make relationships among map elements more visible, heightening one's ability to extract and analyse information.

Two types of data were combined in a GIS to produce a perspective view of a part of San Mateo County, Calif. The digital elevation model, consisting of surface elevations recorded on a 30-meter horizontal grid, shows high elevations as white and low elevations as black. The accompanying Landsat thematic mapper image shows a false-colour infrared image of the same area in 30-meter pixels, or picture elements and combine the two images to produce the three-dimentional image.

Visualization

Maps have traditionally been used to explore the Earth. GIS technology has enhanced the efficiency and analytical power of traditional cartography. As the scientific community recognizes the environmental consequences of human activity, GIS technology is becoming an essential tool in the effort to understand the process of global change. Map and satellite information sources can be combined in models that simulate the interactions of complex natural systems. Through a process known as visualization, a GIS can be used to produce images— not just maps, but drawings, animations, and other cartographic products. These images allow researchers to view their subjects in ways that they never could before. The images often are helpful in conveying the technical concepts of a GIS to nonscientists.

Adding the Element of Time

The condition of the Earth's surface, atmosphere, and subsurface can be examined by feeding satellite data into a GIS. GIS technology gives researchers the ability to examine the variations in Earth processes over days, months, and years. As an example, the changes in vegetation vigor through a growing season can be animated to determine when drought was most extensive in a particular region. The resulting normalized vegetation index represents a rough measure of plant health. Working with two variables over time will allow researchers to detect regional differences in the lag between a decline in rainfall and its effect on vegetation. The satellite sensor used in this analysis is the advanced very high resolution radiometer (AVHRR), which detects the amounts of energy reflected from the Earth's surface at a 1-kilometer resolution twice a day. Other sensors provide spatial resolutions of less than 1 meter.

Serving GIS over the Internet

Through Internet map server technology, spatial data can be accessed and analysed over the Internet. For example, current wildfire perimeters are displayed with a standard web browser, allowing fire managers to better respond to fires while in the field and helping homeowners to take precautionary measures.

The Future of GIS

Environmental studies, geography, geology, planning, business marketing, and other disciplines have benefitted from GIS tools and methods. Together with cartography, remote sensing, global positioning systems, photogrammetry, and geography, the GIS has evolved into a discipline with its own research base known as geographic information sciences. An active GIS market has resulted in lower costs and continual improvements in GIS hardware, software, and data. These developments will lead to a much wider application of the technology throughout government, business, and industry. GIS and related technology will help analyse large datasets, allowing a better understanding of terrestrial processes and human activities to improve economic vitality and environmental quality.

DYNAMICS OF GEOMORPHOLOGY

To place geomorphology upon sound foundations for quantitative research into fundamental principles, it is proposed that geomorphic processes be treated as gravitational or molecular shear stresses acting upon elastic, plastic, or fluid earth materials to produce the characteristic varieties of strain, or failure, that constitute weathering, erosion, transportation and deposition.

Shear stresses affecting earth materials are here divided into two major categories: gravitational and molecular. Gravitational stresses activate all downslope movements of matter, hence include all mass movements, all fluvial and glacial processes. Indirect gravitational stresses activate wave-and tide-induced currents and winds. Phenomena of gravitational shear stresses are subdivided according to behavior of rock, soil, ice, water, and air as elastic or plastic solids and viscous fluids. The order of classification is generally that of decreasing internal resistance to shear and, secondarily, of laminar to turbulent flow.

Molecular stresses are those induced by temperature changes, crystallization and melting, absorption and desiccation, or osmosis. These stresses act in random or unrelated directions with respect to gravity. Surficial creep results from combination of gravitational and molecular stresses on a slope. Chemical processes of solution and acid reaction are considered separately.

A fully dynamic approach requires analysis of geomorphic processes in terms of clearly defined open systems which tend to achieve steady states of operation and are self-regulatory to a large degree. Formulation of mathematical models, both by rational deduction and empirical analysis of observational data, to relate energy, mass, and time is the ultimate goal of the dynamic approach.

Geomorphology

Geomorphology is the scientific study of landforms and the processes that shape them. Geomorphologists seek to understand why landscapes look the way they do, to understand landform history and dynamics, and to predict future changes through a combination of field observations, physical experiments, and numerical modeling. Geomorphology is practiced within geography, geology, geodesy, engineering geology, archaeology, and geotechnical engineering, and this broad base of interest contributes to a wide variety of research styles and interests within the field.

Overview

The surface of Earth is modified by a combination of surface processes that sculpt landscapes and geologic processes that cause tectonic uplift and subsidence. Surface processes comprise the action of water, wind, ice, fire, and living things on the surface of the Earth, along with chemical reactions that form soils and alter material properties, the stability and rate of change of topography under the force of gravity, and other factors, such as (in the very recent past) human alteration of the landscape.

Many of these factors are strongly mediated by climate. Geologic forcings include the uplift of mountain ranges, the growth of volcanoes, isostatic changes in land surface elevation (sometimes in response to surface processes), and the formation of deep sedimentary basins where the surface of Earth drops and is filled with material eroded from other parts of the landscape. The Earth surface and its topography therefore are an intersection of climatic, hydrologic, and biologic action with geologic processes.

The broad-scale topographies of Earth illustrate this intersection of surface and subsurface action. Mountain belts are uplifted due to geologic processes. Denudation of these high uplifted regions produces sediment that is transported and deposited elsewhere within the landscape or off the coast.

On progressively smaller scales, similar ideas apply, where individual landforms evolve in response to the balance of additive processes (uplift and deposition) and subtractive processes (subsidence and erosion).

Often, these processes directly affect each other: ice sheets, water, and sediment are all loads that change topography through flexural isostasy. Topography can modify the local climate, for example through orographic precipitation, which in turn modifies the topography by changing the hydrologic regime in which it evolves. Many geomorphologists are particularly interested in the potential for feedbacks between climate and tectonics mediated by geomorphic processes.

In addition to these broad-scale questions, geomorphologists address issues that are more specific and/or more local. Glacial geomorphologists investigate glacial deposits such as moraines, eskers, and proglacial lakes, as well as glacial erosional features, to build chronologies of both small glaciers and large ice sheets and understand their motions and effects upon the landscape.

Fluvial geomorphologists focus on rivers, how they transport sediment, migrate across the landscape, cut into bedrock, respond to environmental and tectonic changes, and interact with humans.

Soils geomorphologists investigate soil profiles and chemistry to learn about the history of a particular landscape and understand how climate, biota, and rock interact. Other geomorphologists study how hillslopes form and change. Still others investigate the relationships between ecology and geomorphology. Because geomorphology is defined to comprise everything related to the surface of Earth and its modification, it is a broad field with many facets.

Practical applications of geomorphology include hazard assessment (such as landslide prediction and mitigation), river control and stream restoration, and coastal protection.

History

With some notable exceptions, geomorphology is a relatively young science, growing along with interest in other aspects of the Earth Sciences in the mid 19th century. This section provides a very brief outline of some of the major figures and events in its development.

Ancient Geomorphology

Perhaps the earliest one to devise a theory of geomorphology was the polymath Chinese scientist and statesman Shen Kuo (1031-1095 AD). This was based on his observation of marine fossil shells in a geological stratum of a mountain hundreds of miles from the Pacific Ocean. Noticing bivalve shells running in a horizontal span along the cut section of a cliffside, he theorized that the cliff was once the pre-historic location of a seashore that had shifted hundreds of miles over the centuries. He inferred that the land was reshaped and formed by soil erosion of the mountains and by deposition of silt, after observing strange natural erosions of the Taihang Mountains and the Yandang Mountain near Wenzhou. Furthermore, he promoted the theory of gradual climate change over centuries of time once ancient petrified bamboos were found to be preserved underground in the dry, northern climate zone of *Yanzhou*, which is now modern day Yan'an, Shaanxi province.

Early Modern Geomorphology

The first use of the word geomorphology was likely to be in the German language when it appeared in Laumann's 1858 work. Keith Tinkler has suggested that the word came into general use in English, German and French after John Wesley Powell and W. J. McGee used it in the International Geological Conference of 1891.

An early popular geomorphic model was the *geographical cycle* or the *cycle of erosion*, developed by William Morris Davis between 1884 and 1899. The cycle was inspired by theories of uniformitarianism first formulated by James Hutton (1726–1797). Concerning valley forms, uniformitarianism depicted the cycle as a sequence in which a river cuts a valley more and more deeply, but then erosion of side valleys eventually flatten the terrain again, to a lower elevation.

Tectonic uplift could start the cycle over. Many studies in geomorphology in the decades following Davis' development of his theories sought to fit their ideas into this framework for broad scale landscape evolution, and are often today termed "Davisian". Davis' ideas have largely been superseded today, mainly due to their lack of predictive power and qualitative nature, but he remains an extremely important figure in the history of the subject.

In the 1920s, Walther Penck developed an alternative model to Davis', believing that landform evolution was better described as a balance between ongoing processes of uplift and denudation, rather than Davis' single uplift followed by decay. However, due to his relatively young death, disputes with Davis and a lack of English translation of his work his ideas were not widely recognised for many years.

These authors were both attempting to place the study of the evolution of the Earth's surface on a more generalized, globally relevant footing than had existed before. In the earlier parts of the 19th century, authors-especially in Europe-had tended to attribute the form of landscape to local climate, and in particular to the specific effects of glaciation and periglacial processes. In contrast, both Davis and Penck were seeking to emphasize the importance of evolution of landscapes through time and the generality of Earth surface processes across different landscapes under different conditions.

Quantitative Geomorphology

While Penck and Davis and their followers were writing and studying primarily in Western Europe, another, largely separate, school of geomorphology was developed in the United States in the middle years of the 20th century.

Following the early trailblazing work of Grove Karl Gilbert around the turn of the 20th century, a group of natural scientists, geologists and hydraulic engineers including Ralph Alger Bagnold, John Hack, Luna Leopold, Thomas Maddock and Arthur Strahler began to research the form of landscape elements such as rivers and hillslopes by taking systematic, direct, quantitative measurements of aspects of them and investigating the scaling of these measurements. These methods began to allow prediction of the past and future behavior of landscapes from present observations, and were later to develop into what the modern trend of a highly quantitative approach to geomorphic problems. Quantitative geomorphology can involve fluid dynamics and solid mechanics, geomorphometry, laboratory studies, field measurements, theoretical work, and full landscape evolution modeling. These approaches are used to understand weathering and the formation of soils, sediment transport, landscape change, and the interactions between climate, tectonics, erosion, and deposition.

Contemporary Geomorphology

Today, the field of geomorphology encompasses a very wide range of different approaches and interests. Modern researchers aim to draw out quantitative "laws" that govern Earth surface processes, but equally, recognize the uniqueness of each landscape and environment in which these processes operate. Particularly important realizations in contemporary geomorphology include:

1) that not all landscapes can be considered as either "stable" or "perturbed", where this perturbed state is a temporary displacement away from some ideal target form. Instead, dynamic changes of the landscape are now seen as an essential part of their nature.
2) that many geomorphic systems are best understood in terms of the stochasticity of the processes occurring in them, that is, the probability distributions of event magnitudes and return times. This in turn has indicated the importance of chaotic determinism to landscapes, and that landscape properties are best considered statistically. The same processes in the same landscapes does not always lead to the same end results.

Processes

Modern geomorphology focuses on the quantitative analysis of interconnected processes. Modern advances in geochronology, in particular cosmogenic radionuclide dating, optically stimulated luminescence dating and low-temperature thermochronology have enabled us for the first time to measure the rates at which geomorphic processes occur. At the same time, the use of more precise physical measurement techniques, including differential GPS, remotely sensed digital terrain models and laser scanning techniques, have allowed quantification and study of these processes as they happen. Computer simulation and modeling may then be used to test our understanding of how these processes work together and through time.

Geomorphically relevant processes generally fall into (1) the production of regolith by weathering and erosion, the transport of that material, and its eventual deposition. Although there is a general movement of material from uplands to lowlands, erosion, transport, and deposition often occur in closely-spaced tandem all across the landscape.

The nature of the processes investigated by geomorphologists is strongly dependent on the landscape or landform under investigation and the time and length scales of interest.

However, the following non-exhaustive list provides a flavor of the landscape elements associated with some of these.

Primary surface processes responsible for most topographic features include wind, waves, chemical dissolution, mass wasting, groundwater movement, surface water flow, glacial action, tectonism, and volcanism. Other more exotic geomorphic processes might include periglacial (freeze-thaw) processes, salt-mediated action, or extraterrestrial impact.

Fluvial Processes

Rivers and streams are not only conduits of water, but also of sediment. The water, as it flows over the channel bed, is able to mobilize sediment and transport it downstream, either as bed load, suspended load or dissolved load. The rate of sediment transport depends on the availability of sediment itself and on the river's discharge.

Rivers are also capable of eroding into rock and creating new sediment, both from their own beds and also by coupling to the surrounding hillslopes. In this way, rivers are thought of as setting the base level for large scale landscape evolution in nonglacial environments. Rivers are key links in the connectivity of different landscape elements.

As rivers flow across the landscape, they generally increase in size, merging with other rivers. The network of rivers thus formed is a drainage system and is often dendritic, but may adopt other patterns depending on the regional topography and underlying geology.

Aeolian Processes

Aeolian processes pertain to the activity of the winds and more specifically, to the winds' ability to shape the surface of the Earth. Winds may erode, transport, and deposit materials, and are effective agents in regions with sparse vegetation and a large supply of unconsolidated sediments. Although water and mass flow tend to mobilize more material than wind in most environments, aeolian processes are important in arid environments such as deserts.

Hillslope Processes

Soil, regolith, and rock move downslope under the force of gravity via creep, slides, flows, topples, and falls. Such mass wasting occurs on both terrestrial and submarine slopes, and has been observed on Earth, Mars, Venus, Titan and Iapetus. Ongoing hillslope processes can change the topology of the hillslope surface, which in turn can change the rates of those processes. Hillslopes that steepen up to certain critical thresholds are capable of shedding extremely large volumes of material very quickly, making hillslope processes an extremely important element of landscapes in tectonically active areas. On Earth, biological processes such as burrowing or tree throw may play important roles in setting the rates of some hillslope processes.

Glacial Processes

Glaciers, while geographically restricted, are effective agents of landscape change. The gradual movement of ice down a valley causes abrasion and plucking of the underlying rock. Abrasion produces fine sediment, termed glacial flour. The debris transported by the glacier, when the glacier recedes, is termed a moraine. Glacial erosion is responsible for U-shaped valleys, as opposed to the V-shaped valleys of fluvial origin.

The way glacial processes interact with other landscape elements, particularly hillslope and fluvial processes, is an important aspect of Plio-Pleistocene landscape evolution and its sedimentary record in many high mountain environments. Environments that have been relatively recently glaciated but are no longer may still show elevated landscape change rates compared to those that have never been glaciated. Nonglacial geomorphic processes which nevertheless have been conditioned by past glaciation are termed paraglacial processes. This concept contrasts with periglacial processes, which are directly driven by formation or melting of ice or frost.

Tectonic Processes

Tectonic effects on geomorphology can range from scales of millions of years to minutes or less. The effects of tectonics on landscape are heavily dependent on the nature of the underlying bedrock fabric that more less controls what kind of local morphology tectonics can shape. Earthquakes can, in terms of minutes, submerge large extensions creating new wetlands. Isostatic rebound can account for significant changes over thousand or hundreds of years, and allows erosion of a mountain belt to promote further erosion as mass is removed from the chain and the belt uplifts. Long-term plate tectonic dynamics give rise to orogenic belts, large mountain chains with typical lifetimes of many tens of millions of years, which form focal points for high rates of fluvial and hillslope processes and thus long-term sediment production.

Features of deeper mantle dynamics such as plumes and delamination of the lower lithosphere have also been hypothesised to play important roles in the long term (> million year), large scale (thousands of km) evolution of the Earth's topography. Both can promote surface uplift through isostasy as hotter, less dense, mantle rocks displace cooler, denser, mantle rocks at depth in the Earth.

Igneous Processes

Both volcanic (eruptive) and plutonic (intrusive) igneous processes can have important impacts on geomorphology. The action of volcanoes tends to rejuvenize landscapes, covering the old land surface with lava and tephra, releasing pyroclastic material and forcing rivers through new paths. The cones built by eruptions also build substantial new topography, which can be acted upon by other surface processes.

Biological Processes

The interaction of living organisms with landforms, or biogeomorphologic processes, can be of many different forms, and is probably of profound importance for the terrestrial geomorphic system as a whole. Biology can influence very many geomorphic processes, ranging from biogeochemical processes controlling chemical weathering, to the influence of mechanical processes like burrowing and tree throw on soil development, to even controlling global erosion rates through modulation of climate through carbon dioxide balance. Terrestrial landscapes in which the role of biology in mediating surface processes can be definitively excluded are extremely rare, but may hold important information for understanding the geomorphology of other planets, such as Mars.

SCALES IN GEOMORPHOLOGY

Different geomorphological processes dominate at different spatial and temporal scales. Moreover, scales on which processes occur may determine the reactivity or otherwise of landscapes to changes in driving forces such as climate or tectonics. These ideas are key to the study of geomorphology today.

Overlap with other Fields

There is a considerable overlap between geomorphology and other fields. Deposition of material is extremely important in sedimentology. Weathering is the chemical and physical disruption of earth materials in place on exposure to atmospheric or near surface agents, and is typically studied by soil scientists and environmental chemists, but is an essential component of geomorphology because it is what provides the material that can be moved in the first place. Civil and environmental engineers are concerned with erosion and sediment transport, especially related to canals, slope stability (and natural hazards), water quality, coastal environmental management, transport of contaminants, and stream restoration.

Glaciers can cause extensive erosion and deposition in a short period of time, making them extremely important entities in the high latitudes and meaning that they set the conditions in the headwaters of mountain-born streams; glaciology therefore is important in geomorphology.

COMPONENTS OF A GIS

A Geographic Information System combines computer cartography with a database management system. The major components common to a GIS. This diagram suggests that a GIS consists of three subsystems:

(1) an input system that allows for the collection of data to be used and analysed for some purpose;
(2) computer hardware and software systems that store the data, allow for data management and analysis, and can be used to display data manipulations on a computer monitor;
(3) an output system that generates hard copy maps, images, and other types of output.

Two basic types of data are normally entered into a GIS. The first type of data consists of real world phenomena and features that have some kind of spatial dimension. Usually, these data elements are depicted mathematically in the GIS as either points, lines, or polygons that are referenced geographically (or geocoded) to some type of coordinate system. This type data is entered into the GIS by devices like scanners, digitizers, GPS, air photos, and satellite imagery.

The other type of data is sometimes referred to as an attribute. Attributes are pieces of data that are connected or related to the points, lines, or polygons. This attribute data can be analysed to determine patterns of importance. Attribute data is entered directly into a database where it is associated with element data.

Plotted data can be defined as elements because their main purpose is to describe the location of the earthquakes.

For each of the earthquakes, the GIS also has data on their depth. These measurements can be defined as attribute data because they are connected to the plotted earthquake locations shows the attribute earthquake depth organized into three categories: shallow; intermediate; and deep. This analysis indicates a possible relationship between earthquake depth and spatial location-deep earthquakes do not occur at the mid-oceanic ridges.

Within the GIS database a user can enter, analyse, and manipulate data that is associated with some spatial element in the real world. The cartographic software of the GIS enables one to display the geographic information at any scale or projection and as a variety of layers which can be turned on or off. Each layer would show some different aspect of a place on the Earth. These layers could show things like a road network, topography, vegetation cover, streams and water bodies, or the distribution of annual precipitation received.

The difference between element and attribute data can be illustrated. The location of some of the earthquakes that have occurred in the last century. They plotted data points can be defined as elements because their main purpose is to describe the location of the earthquakes.

For each of the earthquakes, the GIS also has data on their depth. These measurements can be defined as attribute data because they are connected to the plotted earthquake locations. The attribute earthquake depth organized into three categories: shallow; intermediate; and deep. This analysis indicates a possible relationship between earthquake depth and spatial location-deep earthquakes do not occur at the mid-oceanic ridges. Within the GIS database a user can enter, analyse, and manipulate data that is associated with some spatial element in the real world. The cartographic software of the GIS enables one to display the geographic information at any scale or projection and as a variety of layers which can be turned on or off. Each layer would show some different aspect of a place on the Earth. These layers could show things like a road network, topography, vegetation cover, streams and water bodies, or the distribution of annual precipitation received. The merges data layers for vegetation community type, glaciers and ice fields, and water bodies (streams, lakes, and ocean).

What is a GIS?

Geographic Information Systems provide a method for integrating and analysing spatial (digital map based) information such as "where is the nearest movie theatre?" alongside related non-spatial information (what movies are playing there?). GIS have three major capabilities (computer mapping, spatial analysis and spatial database) and can operate on a range of platforms (desktop/

laptop computer, Internet, PDA, etc). Many people are becoming far more familiar with seeing the results both textually-for example when their phone shows them the nearest pub-and on open map systems such as Google Maps. Where in the past people had to literally use pencils and string on a paper map to find their nearest school, a computer can do this now extremely quickly an accurately, as long as all the information has been entered correctly in the first place.

In a broader context, GIS involves people and often brings a philosophy of change. For example, in 1994, the New York Police Department introduced GIS to locate crime 'hot-spots', analyse underlying problems and devise strategies and solutions to deal with the problems. Since 1993, violent crime has dropped by two-thirds in New York City. This strategy, known as COMPSTAT, has expanded to cities and jurisdictions across the United States and around the world.

GIS Software

One leading GIS software vendor is ESRI, based in Redlands, California, which offers ArcGIS for the desktop, ArcGIS Server for Internet mapping, ArcPad for PDAs and a range of other products and services for developers. Other popular GIS software packages are available from Cadcorp, Intergraph, MapInfo, Manifold and Autodesk. ERDAS Imagine, ENVI, Idrisi, and PCI Geomatica are geared towards remote sensing i.e analysis of satellite/aircraft images.

There are many third-party extensions and utilities for ArcGIS and other GIS and raster software platforms. Currently, open source GIS software options can be chosen from the first OS GIS package GRASS, recent open source options are DIVA GIS, Quantum GIS, and uDig. There are efforts underway, through the Open GIS Consortium to provide interoperability among spatial data formats and software. The leading contender for spatial data storage is another open source package called PostGIS, which is a spatial extension to the open source database PostgresSQL.

Why is Geospatial Topology Important In A GIS?

Geospatial analysis provides a unique perspective on the world, a tool through which to examine events, patterns, and processes that operate on or near the surface of our planet. To have this tool interact also with the relevant topological information is a must, that is needed to have the complete understanding of the way rivers flow, the weather is affected by the terrain or even how humans and other animals flow or chose to a specific habitat, this kind of information can only be gathered and useful if topological information is present in the GIS.

SPATIAL DATA

Spatial data comes in two major formats, as vector or raster data. The main difference is that a raster is usually a static background picture used to illustrate, whereas a vector is an intelligent ladder of information that can be selected and searched.

Vector (points, lines, polygons)

Vector data often represents anthropogenic (human) features such as roads, buildings, political boundaries (counties, congressional districts, etc), and other features such as lakes and rivers. Vector data is scalable without loss of resolution and is generally represented by XYZ points in a Cartesian frame reference.

Raster (grids/images)

Raster data is pixellated data, and the more pixels that map the data, the better the resolution. However, if raster data is enlarged, it simply enlarges the pixels, which then leads to a loss of resolution. There are many moves being made to make raster data more useable/searchable as it is much faster to collect, unlike vectors where each piece of data usually has a manual input origin.

Raster data is usually derived from satellite imagery or aerial photography (known as remote sensing). Ordinary cameras are only sensitive to visible light. Satellite sensors can capture not only visible light, but also the thermal, microwave, infrared or other types of energy emanating from the Earth's surface. This extra data provides information about sea surface temperatures, vegetation, ozone, etc. Remote sensing is also used to study other planets and extraterrestrial bodies, such as Mars.

GIS TECHNIQUES AND TECHNOLOGY

Modern GIS technologies use digital information, for which various digitized data creation methods are used. The most common method of data creation is digitization, where a hard copy map or survey plan is transferred into a digital medium through the use of a computer-aided design programme, and geo-referencing capabilities.

With the wide availability of ortho-rectified imagery, heads-up digitizing is becoming the main avenue through which geographic data is extracted.

Heads-up digitizing involves the tracing of geographic data directly on top of the aerial imagery instead of by the traditional method of tracing the geographic form on a separate digitizing tablet.

Relating Information from Different Sources

GIS uses spatio-temporal location as the key index variable for all other information. Just as a relational database containing text or numbers can relate many different tables using common key index variables, GIS can relate otherwise unrelated information by using location as the key index variable. The key is the location and/or extent in space-time. Any variable that can be located spatially, and increasingly also temporally, can be referenced using a GIS. Locations or extents in Earth space-time may be recorded as dates/times of occurrence, and x, y, and z coordinates representing, longitude, latitude, and elevation, respectively. These GIS coordinates may represent other quantified systems of temporo-spatial reference. Units applied to recorded temporal-spatial data can vary widely but all Earth-based spatial-temporal location and extent references should, ideally, be relatable to one another and ultimately to a "real" physical location or extent in space-time.

Related by accurate spatial information, an incredible variety of real-world and projected past or future data can be analysed, interpreted and represented to facilitate education and decision making. This key characteristic of GIS has begun to open new avenues of scientific inquiry into behaviours and patterns of previously considered unrelated real-world information.

GIS Uncertainties

GIS accuracy depends upon source data, and how it is encoded to be data referenced. Land Surveyors have been able to provide a high level of positional accuracy utilizing the GPS derived positions. The high-resolution digital terrain and aerial imagery, the powerful computers, Web technology, are changing the quality, utility, and expectations of GIS to serve society on a grand scale, but nevertheless there are other source data that has an impact on the overall GIS accuracy like: paper maps that are not found to be very suitable to achieve the desired accuracy since the aging of maps affects their dimensional stability. In developing a Digital Topographic Data Base for a GIS, topographical maps are the main source of data.

Aerial photography and satellite images are extra sources for collecting data and identifying attributes which can be mapped in layers over a location facsimile of scale. The scale of a map and geographical rendering area representation type are a very important aspects since the information content depends mainly on the scale set and resulting locatability of the map's representations.

In order to digitize a map, the map has to be checked within theoretical dimensions, then scanned into a raster format, and resulting raster data has to be given a theoretical dimension by a rubber sheeting/warping technology process. Uncertainty is a significant problem in designing a GIS because spatial data tend to be used for purposes for which they were never intended. Some maps were made many decades ago, where at that time the computer industry was not even in its perspective establishments. This has led to historical reference maps without common norms. Map accuracy is a relative issue of minor importance in cartography. All maps are established for communication ends. Maps use a historically constrained technology of pen and paper to communicate a view of the world to their users. Cartographers feel little need to communicate information based on accuracy, for when the same map is digitized and input into a GIS, the mode of use often changes. The new uses extend well beyond a determined domain for which the original map was intended and designed. A quantitative analysis of maps brings accuracy issues into focus. The electronic and other equipment used to make measurements for GIS is far more precise than the machines of conventional map analysis. The truth is that all geographical data are inherently inaccurate, and these inaccuracies will propagate through GIS operations in ways that are difficult to predict, yet have goals of conveyance in mind for original design.

Accuracy Standards for 1:24000 Scales Map: 1:24,000 ± 40.00 feet This means that when we see a point or attribute on a map, its "probable" location is within a +/- 40 foot area of its rendered reference, just as to area representations and scale. A GIS can also convert existing digital information, which may not yet be in map form, into forms it can recognize, employ for its data analysis processes, and use in forming mapping output. For example, digital satellite images generated through remote sensing can be analysed to produce a map-like layer of digital information about vegetative covers on land locations. Another fairly recently developed resource for naming GIS location objects is the Getty Thesaurus of Geographic Names which is a structured vocabulary containing about 1,000,000 names and other information about places. Likewise, researched census or hydrological tabular data can be displayed in map-like form, serving as layers of thematic information for forming a GIS map.

Data Representation

GIS data represents real objects with digital data determining the mix. Real objects can be divided into two abstractions: discrete objects and continuous fields. Traditionally, there are two broad methods used to store data in a GIS for both kinds of abstractions mapping references: raster images and vector. Points, lines, and polygons are the stuff of mapped location attribute references. A new hybrid method of storing data is that of identifying point clouds, which combine three-dimensional points with RGB information at each point, returning a "3D colour image". GIS Thematic maps then are becoming more and more realistically visually descriptive of what they set out to show or determine.

Raster

A raster data type is, in essence, any type of digital image represented by reducible and enlargeable grids. Anyone who is familiar with digital photography will recognize the Raster graphics pixel as the smallest individual grid unit building block of an image, usually not readily identified as an artifact shape until an image is produced on a very large scale. A combination of the pixels making up an image colour formation scheme will compose details of an image, as is distinct from the commonly used points, lines, and polygon area location symbols of scalable vector graphics as the basis of the vector model of area attribute rendering.

While a digital image is concerned with its output blending together its grid based details as an identifiable representation of reality, in a photograph or art image transferred into a computer, the raster data type will reflect a digitized abstraction of reality dealt with by grid populating tones or objects, quantities, cojoined or open boundaries, and map relief schemas. Aerial photos are one commonly used form of raster data, with one primary purpose in mind: to display a detailed image on a map area, or for the purposes of rendering its identifiable objects by digitization. Additional raster data sets used by a GIS will contain information regarding elevation, a digital elevation model, or reflectance of a particular wavelength of light, Landsat, or other electromagnetic spectrum indicators. Raster data type consists of rows and columns of cells, with each cell storing a single value.

Raster data can be images with each pixel containing a colour value. Additional values recorded for each cell may be a discrete value, such as land use, a continuous value, such as temperature, or a null value if no data is available. While a raster cell stores a single value, it can be extended by using raster bands to represent RGB colours, colourmaps or an extended attribute table with one row for each unique cell value.

The resolution of the raster data set is its cell width in ground units. Raster data is stored in various formats; from a standard file-based structure of TIF, JPEG, etc. to binary large object data stored directly in a relational database management system similar to other vector-based feature classes. Database storage, when properly indexed, typically allows for quicker retrieval of the raster data but can require storage of millions of significantly sized records.

Vector

In a GIS, geographical features are often expressed as vectors, by considering those features as geometrical shapes.

Different geographical features are expressed by different types of geometry:

- *Points*: Zero-dimensional points are used for geographical features that can best be expressed by a single point reference—in other words, by simple location. Examples include wells, peaks, features of interest, and trailheads. Points convey the least amount of information of these file types. Points can also be used to represent areas when displayed at a small scale. For example, cities on a map of the world might be represented by points rather than polygons. No measurements are possible with point features.
- *Lines or polylines*: One-dimensional lines or polylines are used for linear features such as rivers, roads, railroads, trails, and topographic lines. Again, as with point features, linear features displayed at a small scale will be represented as linear features rather than as a polygon. Line features can measure distance.
- *Polygons*: Two-dimensional polygons are used for geographical features that cover a particular area of the earth's surface. Such features may include lakes, park boundaries, buildings, city boundaries, or land uses. Polygons convey the most amount of information of the file types. Polygon features can measure perimeter and area.

Each of these geometries are linked to a row in a database that describes their attributes. For example, a database that describes lakes may contain a lake's depth, water quality, pollution level.

This information can be used to make a map to describe a particular attribute of the dataset. For example, lakes could be coloured depending on level of pollution. Different geometries can also be compared. For example, the GIS could be used to identify all wells that are within one kilometre of a lake that has a high level of pollution.

Vector features can be made to respect spatial integrity through the application of topology rules such as 'polygons must not overlap'. Vector data can also be used to represent continuously varying phenomena. Contour lines and triangulated irregular networks are used to represent elevation or other continuously changing values. TINs record values at point locations, which are connected by lines to form an irregular mesh of triangles. The face of the triangles represent the terrain surface.

Advantages and Disadvantages

There are some important advantages and disadvantages to using a raster or vector data model to represent reality:

- Raster datasets record a value for all points in the area covered which may require more storage space than representing data in a vector format that can store data only where needed.
- Raster data allows easy implementation of overlay operations, which are more difficult with vector data.
- Vector data can be displayed as vector graphics used on traditional maps, whereas raster data will appear as an image that may have a blocky appearance for object boundaries.
- Vector data can be easier to register, scale, and re-project, which can simplify combining vector layers from different sources.
- Vector data is more compatible with relational database environments, where they can be part of a relational table as a normal column and processed using a multitude of operators.
- Vector file sizes are usually smaller than raster data, which can be 10 to 100 times larger than vector data.
- Vector data is simpler to update and maintain, whereas a raster image will have to be completely reproduced.
- Vector data allows much more analysis capability, especially for "networks" such as roads, power, rail, telecommunications, etc. Raster data will not have all the characteristics of the features it displays.

Non-spatial Data

Additional non-spatial data can also be stored along with the spatial data represented by the coordinates of a vector geometry or the position of a raster cell. In vector data, the additional data contains attributes of the feature. For example, a forest inventory polygon may also have an identifier value and information about tree species. In raster data the cell value can store attribute information, but it can also be used as an identifier that can relate to records in another table.

Software is currently being developed to support spatial and non-spatial decision-making, with the solutions to spatial problems being integrated with solutions to non-spatial problems. The end result with these Flexible Spatial Decision-Making Support Systems is expected to be that non-experts will be able to use GIS, along with spatial criteria, and simply integrate their non-spatial criteria to view solutions to multi-criteria problems. This system is intended to assist decision-making.

Data Capture

Data capture—entering information into the system—consumes much of the time of GIS practitioners. There are a variety of methods used to enter data into a GIS where it is stored in a digital format. Existing data printed on paper or PET film maps can be digitized or scanned to produce digital data. A digitizer produces vector data as an operator traces points, lines, and polygon boundaries from a map. Scanning a map results in raster data that could be further processed to produce vector data. Survey data can be directly entered into a GIS from digital data collection systems on survey instruments using a technique called Coordinate Geometry.

Positions from a Global Navigation Satellite System like Global Positioning System, another survey tool, can also be directly entered into a GIS. Current trend is data collection and field mapping carried out directly with field computers. New technologies allow to create maps as well as analysis directly in the field, projects are more efficient and mapping is more accurate. Remotely sensed data also plays an important role in data collection and consist of sensors attached to a platform. Sensors include cameras, digital scanners and LIDAR, while platforms usually consist of aircraft and satellites.

The majority of digital data currently comes from photo interpretation of aerial photographs. Soft copy workstations are used to digitize features directly from stereo pairs of digital photographs. These systems allow data to be captured in two and three dimensions, with elevations measured directly from a stereo pair using principles of photogrammetry. Currently, analog aerial photos are scanned before being entered into a soft copy system, but as high quality digital cameras become cheaper this step will be skipped.

Satellite remote sensing provides another important source of spatial data. Here satellites use different sensor packages to passively measure the reflectance from parts of the electromagnetic spectrum or radio waves that were sent out from an active sensor such as radar. Remote sensing collects raster data that can be further processed using different bands to identify objects and classes of interest, such as land cover.

When data is captured, the user should consider if the data should be captured with either a relative accuracy or absolute accuracy, since this could not only influence how information will be interpreted but also the cost of data capture. In addition to collecting and entering spatial data, attribute data is also entered into a GIS. For vector data, this includes additional information about the objects represented in the system. After entering data into a GIS, the data usually requires editing, to remove errors, or further processing.

For vector data it must be made "topologically correct" before it can be used for some advanced analysis. For example, in a road network, lines must connect with nodes at an intersection. Errors such as undershoots and overshoots must also be removed. For scanned maps, blemishes on the source map may need to be removed from the resulting raster. For example, a fleck of dirt might connect two lines that should not be connected.

Raster-to-vector Translation

Data restructuring can be performed by a GIS to convert data into different formats. For example, a GIS may be used to convert a satellite image map to a vector structure by generating lines around all cells with the same classification, while determining the cell spatial relationships, such as adjacency or inclusion.

More advanced data processing can occur with image processing, a technique developed in the late 1960s by NASA and the private sector to provide contrast enhancement, false colour rendering and a variety of other techniques including use of two dimensional Fourier transforms. Since digital data is collected and stored in various ways, the two data sources may not be entirely compatible. So a GIS must be able to convert geographic data from one structure to another.

Projections, Coordinate Systems and Registration

A property ownership map and a soils map might show data at different scales. Map information in a GIS must be manipulated so that it registers, or fits, with information gathered from other maps. Before the digital data can be analysed, they may have to undergo other manipulations—projection and coordinate conversions, for example—that integrate them into a GIS.

The earth can be represented by various models, each of which may provide a different set of coordinates for any given point on the Earth's surface. The simplest model is to assume the earth is a perfect sphere. As more measurements of the earth have accumulated, the models of the earth have become more sophisticated and more accurate. In fact, there are models that apply to different areas of the earth to provide increased accuracy. Projection is a fundamental component of map making. A projection is a mathematical means of transferring information from a model of the Earth, which represents a three-dimensional curved surface, to a two-dimensional medium—paper or a computer screen. Different projections are used for different types of maps because each projection particularly suits specific uses.

For example, a projection that accurately represents the shapes of the continents will distort their relative sizes. Since much of the information in a GIS comes from existing maps, a GIS uses the processing power of the computer to transform digital information, gathered from sources with different projections and/or different coordinate systems, to a common projection and coordinate system.

SPATIAL ANALYSIS WITH GIS

Given the vast range of spatial analysis techniques that have been developed over the past half century, any summary or review can only cover the subject to a limited depth. This is a rapidly changing field, and GIS packages are increasingly including analytical tools as standard built-in facilities or as optional toolsets, add-ins or 'analysts'.

In many instances such facilities are provided by the original software suppliers, whilst in other cases facilities have been developed and are provided by third parties. Furthermore, many products offer software development kits, programmeming languages and language support, scripting facilities and/or special interfaces for developing one's own analytical tools or variants.

The website Geospatial Analysis and associated book/ebook attempt to provide a reasonably comprehensive guide to the subject. The impact of these myriad paths to perform spatial analysis create a new dimension to business intelligence termed "spatial intelligence" which, when delivered via intranet, democratizes access to operational sorts not usually privy to this type of information.

2

Geological Image and Geological Texture

Consumers of digital cameras and related products desire high-quality images. Consumer preference for images, however, is not easy to predict. Even if it were technically feasible, creating a perfect reproduction of the light that arrived at the camera would not guarantee the most preferred rendering of the original scene. For example most professional portraiture employs a large degree of image enhancement, and the results are almost always preferred to a veridical rendering. This may occur because most consumers judge the attractiveness of an image without direct reference to the original scene, so that their judgments are based on memory, either of the specific scene or of generic scenes. There is evidence that memory for coloured objects can be unreliable.

Digital images may be modified through the application of image processing algorithms, but what modifications make images look better is not well understood. One approach to this problem is to study directly the effect of image processing on image preference. We recently examined the perceptual performance of demosaicing algorithms in this manner. Previous work has also studied the relation between image colorfulness and human observer quality/naturalness ratings. Here we apply similar experimental methods to study the relation between image tone characteristics and perceptual image quality.

Tone mapping refers to the conversion of input luminance values, as captured by an acquisition device (e.g., a digital camera), to luminance values for display on an output device (e.g., a computer monitor). Luminance values in a natural image can range over about five orders of magnitude. This compares to a much smaller range of about two orders of magnitude available with a computer monitor under typical viewing conditions. Even for the usual situation where the image acquisition device quantizes the number of luminance levels to match the number of levels available on the output device, tone mapping can still improve the appearance of an image. The relation between input and output luminance values produced by a tone-mapping algorithm is called a *tone-mapping curve*.

Tone mapping changes the *tone characteristics* of the image. By tone characteristics we mean the distribution of the luminance values of the image's pixels, without regard to how the pixels are arranged spatially. In general, tone characteristics can either be assessed globally (Over the entire image), or locally (Over some smaller region of interest). Within an image region (either global or local), tone characteristics are completely described by the *luminance histogram* of the region. This specifies the number of image pixels within the region that have each possible output luminance value. In this paper, we will consider both global and local tone characteristics.

Previous work on tone mapping has focused on comparisons of the performance of different tone-mapping methods. Much of this work was conducted in the context of film-based photography, where practical considerations limited attention to global tone-mapping methods in which a single tone-mapping curve was applied to the entire image. Bartleson and Breneman suggested that a good tone-mapping curve established a 1:1 relation between relative perceived brightness values in the scene and the rendered image, where relative brightnesses were computed using a modified power function derived from research on brightness scaling. Their curve corresponded closely to curves that received high ratings in a psychophysical study performed by Clark. Further work by Hunt and co-workers suggested that the Bartleson and Breneman principle should be modified depending on the viewing conditions (in particular the surround of the image) and suggested that although a linear relation between scene and image relative brightnesses was appropriate for reflection prints, a power-law relation between relative brightnesses was more appropriate for transparencies. The widely used zone system for photographic tone mapping (reviewed in Reinhard *et al.*) relies on perceptual judgments of how regions in the original scene appeared to the photographer.

In film photography, it is not practical to automatically adjust the tone-mapping curve between images at separate locations within an image, since the shape of these curves is governed by physical characteristics of the emulsions and film-development process. With the advent of digital imaging, a wider range of tone-mapping algorithms become of practical interest. On the other hand, in many digital cameras image quantization precedes the application of a tone-mapping algorithm, a feature that increases the challenges for successful tone mapping. Thus there has been renewed interest in developing tone-mapping algorithms. Evaluation of these methods has again emphasized comparing the output of competing algorithms.

Algorithms that apply a fixed tone-mapping curve to any image have the feature that the tone characteristics of the images produced by the algorithm can vary widely, since these characteristics depend strongly on the input. Digital imaging presents the opportunity to develop algorithms using a different principle. Rather than defining the relationship between input and output luminances, one can specify target output tone characteristics and apply an image-dependent transformation that yields a good approximation of these characteristics. One early digital tone-mapping algorithm, histogram equalization, is based on this idea: the algorithm maps the luminance values in the input image to produce a desired luminance histogram in the output image. Although it seems unlikely that the optimal output histogram is completely independent of image content, the principle of specifying target output image tone characteristics has been incorporated into recent tone-mapping algorithms intended to improve upon histogram equalization. In these algorithms, the output histogram varies with an analysis of image content.

The work we present here is intended to further explore the idea that effective tone mapping can be achieved through specification of desired output image tone characteristics. Rather than focusing on the development and evaluation of tone-mapping algorithms, we chose to address the underlying issue of whether we could identify output tone characteristics that produce perceptually attractive images, and whether such characteristics depend on image content. To this end, we report the results of two image preference studies and analyze how image preference is related to image tone characteristics.

The work presented here employs images captured with standard digital cameras and is directed at improving the quality of images produced from such cameras. We do not explicitly consider the case where the dynamic range of the capture and display devices varies greatly.

As most amateur digital photographs include people, our studies employ an image set that consisted mainly of images of people. We also wanted to include images of people from different ethnic backgrounds, since many earlier tone-mapping studies used images of Caucasians only.

Experiment 1 was exploratory, with the goal of identifying systematic relationships between tone variables and image quality. We applied four different tone-mapping methods to each of 25 experimental images and measured the perceptual quality of the different renderings of each image. These algorithms produced output images with a range of tone characteristics. Image preference was measured using a pairwise comparison procedure. On each trial, observers indicated which of two presented images was the most attractive.

The pairwise comparison procedure is intuitive for observers and yields reliable data. Note, however, that observers only make judgments about different renderings of the same input image. Thus some analysis is required to aggregate a data set large enough to explore the question of how an image's tone characteristics relate to its perceptual quality. To this end, the preference choice data were analyzed using a regression procedure to yield metric differences in image quality between image pairs. The procedure yields difference ratings that are commensurate across input images. We then asked whether differences in specific image *tone variables* were predictive of the difference ratings. Here the term tone variable refers to a summary measure, such as mean luminance, that may be computed from the output luminance histogram.

We used 25 digitally acquired images and rendered each on a CRT computer monitor using four different tone-mapping methods. The four methods produced results that were perceptually different for most of the images, thus providing variation in image tone characteristics whose effect we could study.

Methods: Image Acquisition: Twenty-five images were used in Experiment 1. Twenty-one were captured in Santa Barbara, California and four were taken in Palo Alto, California. All of the images were captured under daylight, at different times of the day, throughout May 1999. The illuminant was measured immediately following the acquisition of each image by placing a white reflectance standard in the scene and measuring the reflected light using a Photo Research PR-650 spectraradiometer. Of the 25 images, 17 were portraits of people, 5 were landscapes, and 3 were of objects.

The 21 Santa Barbara images were taken with a Kodak DCS-200 camera and the 4 Palo Alto images with a Kodak DCS-420 camera. Both cameras have a resolution of 1524×1012 with RGB sensors arranged in a Bayer mosaic. The DCS-200 captures the input light intensity using 8-bit linear quantization, whereas the DCS-420 captures with 12-bit precision. The 12-bit values captured by the DCS-420 are converted to 8-bit values on-camera via a nonlinear transformation. The relative RGB spectral sensitivities and response properties of both cameras were characterized as described elsewhere. This characterization left one free parameter describing the overall sensitivity of the camera undetermined, as this parameter varies with acquisition exposure duration and f-stop. The images were cropped to a maximum size of 575 (w) by 800 (h) pixels to ensure that two renderings of each image could be displayed simultaneously on the computer screen used in our experiment.

Image Processing: For the DCS-200, a dark level was subtracted from the raw quantized pixel values before further processing. The dark level was estimated from an image acquired with the lens cap on and computing the spatial average of the resulting image.

The average for the red, green, and blue sensors were all 13.5 on the camera's 8-bit (0–255) output scale and this is the value that was subtracted. To estimate the dynamic range of the images, we compared the minimum and maximum pixel values for the green sensor. These typically occupied the entire allowable output range (approximately 13–255 before dark subtraction). Given that some pixels had values near zero after dark subtraction, it is not possible to express the dynamic range of these images as a meaningful ratio.

For the DCS-420, it was possible to linearize the output values using a look-up table provided as part of each raw image file. This was done prior to further processing. After linearization, the estimated dark level for the DCS-420 was close to zero and no explicit dark level subtraction was performed. The dynamic range of these images could be estimated by taking the ratio of the maximum to minimum linearized output value for the green sensor. These ratios varied from 17 to 140 across the DCS-420 images used in this experiment.

Demosaicing : Because the two cameras employed a mosaiced design, with each raw pixel corresponding to only one of the three sensor types, it was necessary to apply a demosiacing algorithm to convert the raw mosaiced image to a full color RGB image. We used a Bayesian demosaicing algorithm developed by Brainard and colleague and summarized in a recent paper where we evaluated the perceptual quality of demosaicing algorithms. (The performance of the Bayesian algorithm is controlled by a number of parameters. For the application here, the correlation between nearest-neighbor pixels was assumed to be 0.90, whereas the correlation between the responses of different sensor classes at the same image location was estimated from a bilinear interpolation of the mosaiced image. Finally, the algorithm assumed that there was additive normally distributed pixel noise with a standard deviation for each sensor class equal to 4% of the spatial average of responses for that class. The estimates at each location were obtained by applying the algorithm to a 5×5 image region surrounding that pixel.) The demosaicing results for our images were in general quite good, with very few noticeable artifacts.

Color balancing : We by-passed the on-board color balancing of the cameras and used our measurements of the scene illuminants to color balance the images. Given the camera's RGB sensor relative spectral sensitivities and the measured illuminant, we were able to estimate the relative surface spectral reflectance of the object at each scene location. This was done using a Bayesian estimation procedure that will be described in a future report. Briefly, we constructed a normally distributed multivariate prior distribution for object surface reflectances by analyzing the Vrehl *et al.* data set of measured surface reflectance functions. The analysis followed closely the method introduced by Brainard and Freeman in their work on computational color constancy. Given the prior, estimating reflectances from the sensor responses is a straightforward application of Bayes rule. Using the estimated surface reflectance functions, we could then synthesize an image that consisted of the CIE XYZ tristimulus coordinates that would have been obtained had the surface been viewed under standard CIE daylight D65, up to an overall scale factor. This scale factor varied from image to image depending on the scene illuminant, acquisition exposure, and acquisition f-stop. Uncertainty about the scale factor is equivalent to uncertainty about the overall intensity of the scene illuminant and is thus handled transparently by the tone-mapping algorithms that we applied to render the images, which are designed to apply to images captured over a wide range of overall scene luminances. Note that image L* properties reported in this paper refer to L* values for the experimental images displayed on the experimental monitor, not to L* properties of regions of the original scene.

To check the accuracy of the color balancing process, an image of a Macbeth color checker was taken using the Kodak DCS-420 digital camera. Raw RGB values (before demosaicing) were extracted for each of the 24 color checker patches. The Bayes color correction was used to estimate the XYZ values of the patches under CIE D65 illumination. These estimates were compared with target values computed from measured spectral reflectances of the color checker patches and the known spectral relative spectral power distribution of CIE daylight D65. Here the free overall scale factor was determined so that the two middle gray color checker patches (patch Nos. 21 and 22) matched in average luminance between the color balanced and target values. The average CIELAB ÄE 94 difference between the estimated values and directly determined target values (average taken over the 24 patches) was 3.6 units, indicating that the algorithm worked well.

Tone mapping : Four tone-mapping algorithms (*Clipping, Histogram Equalization, Larsons Method, and Holms Method*) were applied to the color balanced XYZ images. These are described below. Each method transformed the luminance of each image pixel while holding the chromaticity of each pixel constant. The relation between a particular measurement of input and output luminance is referred to as the algorithms *tone-mapping curve*. In general, the tone-mapping curve produced by an algorithm depends on image content. Each of the algorithms used was global, in the sense that the same tone-mapping curve was applied to every pixel in the image. It should be emphasized that our main goal was to use a variety of tone-mapping algorithms that would produce different tone-mapping characteristics. The performance of each algorithm was not of primary concern. All four methods led to acceptable (as judged by the authors) renderings for all of the images.

Clipping : For the clipping method, the tone-mapping curve relating image luminance to display luminance was a straight line through the origin. Image luminances that were mapped to display luminances greater than the maximum available on the output device were clipped to the maximum. The slope of the tone-mapping curve was determined so that maximum display luminance was equal to five times the mean luminance of the tone-mapped image. This clipping method provides a simple baseline that works reasonably well.

Histogram equalization : A widely used method that re-assigns luminance values to achieve a particular target luminance histogram (e.g., uniform or Gaussian) in the tone-mapped image. This method efficiently uses the dynamic range of the display device, but can generate images that have exaggerated contrast and thus a harsh appearance. In our implementation, the target histogram was a Gaussian centered at the middle of the output range.

Larson method : A more sophisticated version of histogram equalization. The idea is to limit the magnitude of luminance mapping, so that luminance differences within the image that were not visible before tone mapping are not made visible by it. Images tone mapped with the Larson method generally have a more natural appearance than when using the traditional histogram equalization method.

Holms method : Part of a color reproduction pipeline created at Hewlett-Packard Labs for use in digital cameras. We used only the tone-mapping segment of the pipeline for consistency with the other methods. In Holms method, the input image is first classified as one of several different types (e.g., high key or low key) using a set of image statistics. A tone-mapping curve is then generated according to the image type and image statistics, and this curve is applied to the whole image. This method incorporates preference guidelines that came from the inventors extensive experience in photographic imaging.

Rendering for display : The images were presented on a CRT monitor. Conversion between the tone-mapped XYZ values and monitor settings was achieved using the general model of monitor performance and calibration procedures described by Brainard. The calibration was performed using the PR-650 spectraradiometer. Spectral measurements were made at 4 nm increments between 380 and 780 nm but interpolated with a cubic spline to the CIE recommended wavelength sampling of 5 nm increments between 380 and 780 nm. CIE XYZ coordinates were computed with respect to the CIE 1931 color matching functions.

ROOM AND DISPLAY SETUP

The experimental room was set up according to the International Organization for Standardization Recommendations for Viewing Conditions for Graphic Technology and Photography. The walls were made of a medium gray material and the table on which the monitor was placed was covered with black cloth. The room was lit by two fluorescent ceiling lights (3500 K) controlled by a dimmer switch set at a dim level. The illumination measured at the observer position was 41 lux. The experiment was controlled by MATLAB software based on the Psychophysics Toolbox. The images were displayed on a Hewlett Packard P1100 21 in. monitor (1280×1024 pixels) driven by a Hewlett Packard Kayak XU computer.

Procedure : On each trial of the experiment, observers were shown pairs of the same scene rendered via different tone-mapping methods and were asked to choose the image that they found to be the most attractive. To further explain this instruction, observers were asked to choose the image they would select to put into their own photo album. The observers were also asked to look around the images before making a decision rather than focus on just one aspect.

The experiment started after a 2 min adaptation period. Three seconds after each pair of images was presented, two selection boxes appeared under the images. This 3 s delay was to encourage the observers to carefully consider their decision. There was no upper limit on response time. The observers indicated their preference by using a mouse to move a cursor to the selection box under the preferred image and clicking. The observer could subsequently change his/her mind by clicking on the alternative box. When the observer was satisfied with his/her selection, he/she clicked on an enter button to move to the next trial. The images were viewed from a distance of 60 cm. The images ranged in width from 17 to 19 cm (subtending visual angles 16.1° to 18.0°) and ranged in height from 13 to 25 cm (subtending visual angles from 12.4° to 23.5°). Images were shown in pairs on the monitor, one on the left and one on the right. Each image had a border of width 1 cm which was rendered as the brightest simulated D65 illuminant the monitor could produce (78 cd/m2). The remaining area of the monitor emitted simulated D65 illuminant but at a luminance level of about 20% of the border region (measured at 14.9 cd/m2). Using four rendering methods gives six pairwise presentation combinations per image. For the 25 experimental images, this produces a stimulus set of 150 image pairs.

Observers : Twenty observers participated in the experiment (12 males and 8 females) with an average age of 31 (range 19–62). The experiment took place at Hewlett Packard Labs in Palo Alto and the observers were recruited by posting flyers around the building complex. The observers were a mixture of Hewlett Packard employees and outside friends and family. Only color normal observers participated. Color vision was tested using the Ishihara color plates.

Data analysis : The aim of our analysis was to summarize image tone characteristics using simple tone variables, and to determine whether these variables predicted image preference. We hoped to identify systematic relationships between preference ratings and tone variables. Thus our data analysis has two important components: the procedure used to transform the pairwise image judgments to image preference ratings and the procedures used to extract variables that capture image tone-mapping characteristics.

TOOLS

There are two primary tools used in the field to conduct terrain/battlefield evaluation and visualization. The Digital Topographic Support System (DTSS) is the terrain team's primary automated tool used to create terrain products.

TerraBase II is the primary automated PC-based tool used by units at the brigade level and below. Each tool has advantages and limitations, and both are used to give planners optimal battlefield evaluation and visualization.

The DTSS is a set of automation equipment consisting of a powerful CPU, a 24" or 36" roll/drum scanner and several 36" color plotters, although the actual amounts and numbers of equipment on hand will vary between units. The system is generally mounted on a HMMWV, 5-ton, FMTV or other mobile platform. Until recently, terrain teams were only present at the division and corps levels and provided support to their command group and MSCs, so the DTSS was only available at the division level or higher. However, under the Force XXI restructuring initiatives, there is now a DTSS suite and a four-person terrain detachment with each maneuver brigade in the digital corps.

The DTSS is invaluable when units are preparing to enter an area of operations. It is a powerful tool that can create extremely precise products that evaluate all relevant considerations: terrain, weather, soils, hydrology, vegetation, and other factors. It can create products such as soil analysis and vegetation overlays and a MCOO. These products are usually made well before a unit enters a particular AO. The advantage of the DTSS is that all the input data and output products are digital and can be stored, sent, printed, or later manipulated to fit any particular need. The system also has a limited rapid reproduction capability so that products can be made and printed while in theater. More leaders need to realize the capabilities of this system and use its products in support of the continuous Military Decision-Making Process/ Intelligence Preparation of the Battlefield (MDMP/IPB) processes.

Terrain teams normally produce a general suite of products for each mission to support the MDMP. These include a MCOO, separate specialized cross-country mobility, hydrology, lines of communication, elevation/slope tint and soils products. These products are often distributed down to the brigade S-2 in hardcopy format, although local distribution SOP may vary. These products are intended for use by all BOSs; all planners need to know these products exist and ensure they have access to them as soon as they are available. In addition to this general set of products, teams and detachments can provide more detailed and customized products, depending on the situation and request.

There are several different programs directly available to the unit-level planner to conduct terrain analysis tasks, but the most commonly used and available is TerraBase II. TerraBase II displays terrain in 3D perspective views and conducts relief-based analysis. Many types of imagery and digital maps can be draped over a 3D relief of the terrain to provide a very useful representation of the terrain features.

TerraBase II is a freeware program created to support units at the brigade level and below. It is an easy-to-use terrain evaluation tool usable on low-end PCs. The program provides planners from any BOS the ability to create the specific products important to the ground soldier, such as lines of sight, weapons fans and image maps. The Engineer School currently uses TerraBase in engineer officer training, to support the expectation of maneuver commanders that the engineer is the terrain expert.

The program uses National Imagery and Mapping Agency (NIMA) data, United States Geological Survey (USGS) data and many other types of data, including satellite imagery.

The main functions planners use are the weapons fan/range circle and line-of-sight tools. Some additional analysis functions include slope and elevation analyses, terrain categorization, and oblique and perspective views. The program also creates "movies" showing a rotating panoramic view from a specified location, or "fly-throughs" over the terrain using user-identified viewpoint information. The program is limited in that it does not consider the effects of vegetation, hydrology or weather, so the planner must always evaluate the validity and utility of the TerraBase II products based on these unassessed criteria.

Planners can download versions 3.0, 4.0, and 5.0 of the program, training documents including videos on how to execute functions, and a Microsoft Word tutorial from the Internet. Version 3.0 provided the basis for a multimedia CD-ROM TerraBase II tutorial package recently distributed by the U.S. Army Engineer School to each active duty, USAR and ARNG engineer battalion, brigade and group, to each of the tactical branch schools and to the CTCs.

The intent is to provide the field a means to self-train on the program using the package plus any locally reproduced copies of the package. The downloaded programs come with very little digital data, but the tutorial package includes digital elevation, map and imagery data for common military training areas, such as NTC, JRTC and CMTC, plus data for other areas of interest. The CD-ROM is currently given to each student at the engineer basic, advanced and pre-command courses, and once annually to engineer students at CGSC.

The program can display high-resolution imagery, including 1-meter resolution products from NIMA, the USGS or commercial sources. The program displays GeoTIFF images created using high-resolution (1m) color MrSID (Multiresolution Seamless Image Database) imagery. Planners can request this imagery for Fort Irwin, CA, Fort Polk, LA, and Fort Leonard Wood, MO, and use it to create image files usable in TerraBase II. This imagery allows the user to combine the tactical analysis tools of TerraBase with satellite imagery that commanders and staffs often find easier to interpret and more useful than digital map data. Other program uses include, but are not limited to:

- Planning direct fire on/from proposed battle positions.
- Re-transmission position planning.
- Confirming visibility to objectives/NAIs/TAIs.

- Estimating slope along routes for resupply vehicles and others.
- Delineation of air avenues of approach.
- Identifying optimal locations for LZs/DZs/base camps.
- Placing OPs for optimal viewing and concealed movement.
- Placing artillery pieces on low-slope areas.
- Identifying potential ambush sites.
- Conducting 3D movement/mission rehearsals.

TerraBase is free and easy to learn, and anyone in the force can use and apply TerraBase and its products to their missions. However, it is important to know that much of the digital image, map and elevation data used is "Limited Distribution" and only DOD military and approved civilian personnel are authorized to use the data.

The planner must make or request TerraBase II/terrain team products before, during and after the mission. Products received before the mission assist in planning. Products made or received during the mission assist in course correction and mission flexibility. Products made after the mission assist in BDA, analysis of the effectiveness of the maneuver and the breach, and augment after-actions/lessons learned records.

Products

There are three basic types of topographic or terrain visualization products with which the planner needs to be familiar: standard, nonstandard and TerraBase II products.

Standard products are hardcopy or softcopy items which are mass-produced for the DOD and are available through the supply system using an NSN. These include hardcopy or digital data such as maps, digital imagery, elevation data, and other feature data. The hardcopy products are normally paper, while digital products are normally on CD-ROMs.

Nonstandard products also include hardcopy and softcopy items, but they are not mass-produced. Terrain teams create these products using the DTSS and they are tailored to support specific operations. These products graphically illustrate the latest information about an area and serve as tactical decision aids (TDAs). A TDA is a product used to assist planners and commanders in making decisions about operations based on the effects of terrain and weather. An example is a cross-country mobility model illustrating the speed at which a mechanized brigade can move off-road. A day-old satellite image with military grid lines and a legend is much more useful than a 35-year old map.

Some other TDAs planners could use include Terrain Category (TerCat) and elevation tint products, which BOSs, such as Field Artillery and Air Defense Artillery, often use.

Planners can use TerCat products to identify low-slope areas from which artillery pieces can fire, while elevation tints are used to identify air AAs. A sample package of nonstandard TDA products created for an operation could include the items listed below; the full range of products is listed in FM 3-34.230:

- Observation and fields of fire.
- Concealment.
- MCOO.
- Key terrain.
- Avenues of approach.
- LOC-roads, bridges, tunnels, sharp curves, steep grades.
- Hydrology-depth, width, velocity, bottom material, bank height, ford sites.
- Facilities-natural resources, manufacturing facilities.
- Image map (may be combined with facilities and key terrain).
- Elevation tint, placed over maps or imagery.
- LZ/DZ.

The third type of product is the TerraBase product created at the unit level. This is simply any product created using any function of TerraBase which is then used for tactical terrain visualization, evaluation and decisionmaking.

A U.S. Army Engineer School TTP tool for identifying and collating all the various products necessary or desired for the terrain planner is called the Terrain Mission Folder (TMF) (mentioned in both FM 3-34.2 and FM 3-34.230). This product is a folder or collection of terrain/weather data necessary to support the planner in the decisionmaking process. The TMF is a set of mission-specific topographic products that is focused on a 10km x 10km area of operations, and is created in advance of the operation to support the IPB process. It may include:

- Standard products, either hardcopy or digital.
- Nonstandard products created by a terrain team.
- TerraBase II products specifically created for the operation.
- Intelligence and weather reports.
- Graphics from the Engineer planner.

- Photographs.
- Any other data deemed important to the planner.

TRAINING

Knowledge of the process of terrain, weather and situational analysis and how to leverage the various terrain tools, organizations and products is key to the planner's ability to make the correct recommendation to the commander. Therefore, staff planners must ensure they and their units receive training across the full spectrum of the terrain and weather integration process. The training should occur at home station, and should focus on several areas of interest. This home-station training should include:

- Periodic refreshers on the MDMP and IPB processes, with special focus on terrain analysis. Training should always include a full review and integration of terrain tools, products and area evaluations.
- Unit-or individual-level training on the TerraBase II program, including practical exercises on selected areas of interest. This allows the exercise products to remain at the unit for potential later use during deployments/field training. Where and when possible, each product should be field-tested for validity and this knowledge incorporated into future products.
- The planner should request that the terrain team warrant officer (OIC) brief the unit and demonstrate/show the products the team can provide the planners, when and how the planner should request them and when each product is likely to be available during the planning process. It is important to use a sample of the actual products during this training.
- Staff-exercise coordination should be a part of all training and exercises to ensure information and product flow does not stop at any level or echelon but makes it down to the bottom, user level. The staff-exercise should also ensure each BOS planner incorporates his specialty into the process.

Terrain Analysis

Whether concrete, sand, snow, or jungle, knowing your terrain and the obstacles you may face on a mission can mean all the difference in completing your goals. Terrain analysis software from Intergraph delivers extensive topographical and strategic information. From terrain detection to terrain visualization, Intergraph terrain analysis solutions provide preparatory support for mission planning systems and beyond.

Whether preparing for an environmental engineering project, creating emergency response simulation, or devising new supply delivery route options, terrain analysis tools from Intergraph provide the flexibility and power that your military or intelligence agency needs.

- Terrain Analysis Software: From terrain detection to terrain visualization and more, learn more today about how our software solutions work together to deliver comprehensive terrain analysis tools.
- Contact Intergraph: If you are interested in learning more about the specific ways Intergraph terrain analysis software can benefit your enterprise, please contact us via our information request form today. Additionally, to stay current on technological advancements in terrain analysis, terrain detection, terrain visualization, and more, simply sign up for our e-Connection newsletter.

Terrain Analysis Solutions

As terrain differs depending on geography and objectives, Intergraph offers a wide selection of terrain analysis solutions that allow your military or intelligence agency to learn what obstacles you may need to go around, over, or through to reach your desired outcomes.

Proper terrain analysis comes with the ability to present a clear picture of topographical, elevation, and relief parameters. Intergraph provides the solutions your organization needs to meet all of your terrain analysis requirements.

Terrain Analysis Software from Intergraph

Whether being used as part of a military planning system, for humanitarian relief missions, or for environmental engineering planning, terrain analysis software from Intergraph provides an extensive view of the terrain involved as a map or as a 3D interactive model for unparallel degrees of flexibility.

EVALUATION OF GROUNDWATER POTENTIAL

In view of increasing demand of water for various purposes like agricultural, domestic, industrial etc., a greater emphasis is being laid for a planned and optimal utilization of water resources. Due to uneven distribution of rainfall both in time and space, the surface water resources are unevenly distributed. Also, increasing intensities of irrigation from surface water alone may result in alarming rise of water table creating problems of waterlogging and salinization, affecting crop growth adversely and rendering large areas unproductive. This has resulted in increased emphasis on development of ground water resources. The simultaneous development of ground water specially through dug wells and shallow tubewells will lower water table, provide vertical drainage and thus can prevent waterlogging and salinization.

Areas which are already waterlogged can also be reclaimed. On the other hand continuous increased withdrawals from a ground water reservoir in excess of replenishable recharge may result in regular lowering of water table. In such a situation, a serious problem is created resulting in drying of shallow wells and increase in pumping head for deeper wells and tubewells. This has led to emphasis on planned and optimal development of water resources.

An appropriate strategy will be to develop water resources with planning based on conjunctive use of surface water and ground water.

For this the first task would be to make a realistic assessment of the surface water and ground water resources and then plan their use in such a way that full crop water requirements are met and there is neither waterlogging nor excessive lowering of ground water table.

It is necessary to maintain the ground water reservoir in a state of dynamic equilibrium over a period of time and the water level fluctuations have to be kept within a particular range over the Water balance techniques have been extensively used to make quantitative estimates of water resources and the impact of man's activities on the hydrologic cycle. The study of water balance is defined as the systematic presentation of data on the supply and use of water within a geographic region for a specified period. With water balance approach, it is possible to evaluate quantitatively individual contribution of sources of water in the system, over different time periods, and to establish the degree of variation in water regime due to changes in components of the system.

The basic concept of water balance is :

Input to the system-outflow from the system = change in storage of the system (over a period of time)

The general methods of computations of water balance include:

(i) identification of significant components,

(ii) evaluating and quantifying individual components, and

(iii) presentation in the form of water balance equation.

Ground Water Balance Equation

Considering the various inflow and outflow components, the terms of the ground water balance equation can be written as:

Ri + Rc + Rr + Rt + Si + Ig = Et + Tp + Se + Og +.S

where,

Ri = recharge from rainfall;

Rc = recharge from canal seepage;

Rr = recharge from field irrigation;

Rt = recharge from tanks;

Si = influent seepage from rivers;

Ig = inflow from other basins;

Et = evapotranspiration;

Tp = draft from ground water;

Se = effluent seepage to rivers;

Og = outflow to other basins; and

.S = change in ground water storage.

This equation considers only one aquifer system and thus does not account for the interflows between the aquifers in a multi-aquifer system. However, if sufficient data related to water table and piezometric head fluctuations and conductivity of intervening layers are available, the additional terms for these interflows can be included in the governing equation.

All elements of the water balance equation are computed using independent methods wherever possible. Computations of water balance elements always involve errors, due to shortcomings in the techniques used. The water balance equation therefore usually does not balance, even if all its components are computed by independent methods.

The discrepancy of water balance is given as a residual term of the water balance equation and includes the errors in the determination of the components and the values of components which are not taken into account.

The water balance may be computed for any time interval. The complexity of the computation of the water balance tends to increase with increase in area. This is due to a related increase in the technical difficulty of accurately computing the numerous important water balance components.

Study Area

A basinwise approach yields the best results where the ground water basin can be characterized by prominent drainages. A thorough study of the topography, geology and aquifer conditions should be taken up. The limit of the ground water basin is controlled not only by topography but also by the disposition, structure and permeability of rocks and the configuration of the water table. Generally, in igneous and metamorphic rocks, the surface water basin and ground water basin are coincident for all practical purposes, but marked differences may be encountered in stratified sedimentary formations. Therefore, the study area for ground water balance study is preferably taken as a doab which is bounded on two sides by two streams and on the other two sides by other aquifers or extension of the same aquifer.

Study Period

In areas where most of the rainfall occurs in a part of year, it is desirable to conduct water balance study on part year basis, that is, for monsoon period and non-monsoon period.

Generally, the periods for study in such situations will be from the time of maximum water table elevation to the time of minimum water table elevation as the non-monsoon period and from the time of minimum water table to the time of maximum water table elevation as monsoon period.

For northern India, the water year can be taken as November 1 to October 31 next year. The monsoon and non-monsoon periods can be taken as June to October and November to May next year respectively. It is desirable to use the data of a number of years preferably covering one cycle of a dry and a wet year.

Data Requirement

The data required for carrying out the ground water balance study can be enumerated as follows : Rainfall data : Monthly rainfall data of sufficient number of stations lying within or around the study area should be available. The location of raingauges should be marked on a map.

Land use data and cropping patterns : Land use data are required for estimating the evapotranspiration losses from the water table through forested area. Crop data are necessary for estimating the spatial and temporal distributions of the ground water withdrawals and canal releases, if required. Evapotranspiration data and monthly pan evaporation rates should also be available at few locations for estimation of consumptive use requirements of different crops.

River data : River data are required for estimating the interflows between the aquifer and hydraulically connected rivers. The data required for these computations are the river gauge data, monthly flows and the river cross-sections at a few locations.

Canal data : Monthwise releases into the canal and its distributories along with running days each month will be required. To account for the seepage losses, the seepage loss test data will be required in different canal reaches and distributories.

Tank data : Monthly tank gauges and releases should be available. In addition to this, depth vs area and depth vs capacity curves should also be available. These will be required for computing the evaporation and the seepage losses from tanks. Also field test data will be required for computing final infiltration capacity to be used to evaluate the recharge from depression storage.

Aquifer parameters : The specific yield and transmissivity data should be available at sufficient number of points to account for the variation of these parameters within the area.

Water table data : Monthly water table data or at least premonsoon and post-monsoon data of sufficient number of wells should be available. The well locations should be marked on a map. The wells should be adequate in number and well distributed within the area, so as to permit reasonably accurate interpolation for contour plotting. The available data should comprise reduced level (R.L.) of water table and depth to water table.

Draft from wells : A complete inventory of the wells operating in the area, their running hours each month and discharge are required for estimating ground water withdrawals. If draft from wells is not known, this can be obtained by carrying out sample surveys.

Estimation of Ground Water Balance Components

The estimation of the various inflow and outflow components of the ground water balance equation are discussed below.

Recharge from Rainfall (Ri): Part of the rain water, that falls on the ground, is infiltrated into the soil. This infiltrated water is utilized partly in filling the soil moisture deficiency and part of it is percolated down reaching the water table.

This water reaching the water table is known as the recharge from rainfall to the aquifer. Recharge due to rainfall depends on various hydrometeorological and topographic factors, soil characteristics and depth to water table. The methods for estimation of rainfall recharge involve the empirical relationships established between recharge and rainfall developed for different regions, ground water estimation committee norms, water balance approach, and soil moisture data based methods.

Empirical Methods

Chaturvedi Formula : Based on the water level fluctuations and rainfall amounts in Ganga-Yamuna doab, Chaturvedi in 1936, derived an empirical relationship to arrive at the recharge as a function of annual precipitation.

R = 2.0 (P-15)0.4

where,

R = net recharge due to precipitation during the year, in inches;

P = annual precipitation, in inches.

This formula was later modified by further work at the U. P. Irrigation Research Institute, Roorkee and the modified form of the formula is

R = 1.35 (P-14)0.5

The Chaturvedi formula has been widely used for preliminary estimations of ground water recharge due to rainfall.

Amritsar Formula : Using regression analysis for certain doabs in Punjab, Sehgal developed a formula in 1973 for Irrigation and Power Research Institute, Punjab.

R = 2.5 (P-16)0.5

where, R and P both are measured in inches.

Krishna Rao : Krishna Rao gave the following empirical relationship in 1970 to determine the ground water recharge in limited climatological homogeneous areas :

R = K (P-X)

The following relation is stated to hold good for different parts of Karnataka :

R = 0.20 (P-400) for areas with annual normal rainfall (P) between 400 and 600 mm

R = 0.25 (P-400) for areas with P between 600 and 1000 mm

R = 0.35 (P-600) for areas with P above 2000 mm where, R and P are expressed in millimeters.

The relationships indicated above, which were tentatively proposed for specific hydrogeological conditions, have to be examined and established or suitably altered for application to other areas.

Ground Water Estimation Committee Norms

A committee 'Ground Water Estimation Committee' was constituted in 1982 to improve the existing methodologies for estimation of the ground water resource potential. It was recommended that the ground water recharge should be estimated based on ground water level fluctuation method. However, in areas, where ground water level monitoring is not being done regularly, or where adequate data about ground water level fluctuation is not available, adhoc norms of rainfall infiltration may be adopted. As a guideline, following norms for recharge from rainfall may be adopted :

(i) Alluvial areas

- In sandy areas-20 to 25 percent of rainfall
- In areas with higher clay content-10 to 20 percent of rainfall

(ii) Semi-consolidated sandstones

- Friable and highly porous-10 to 15 percent of rainfall

(iii) Hard rock areas.

Granitic terrain :

- Weathered and fractured-10 to 15 percent of rainfall Unweathered-5 to 10 percent of rainfall.

Basaltic terrain :

- Vesicular and jointed basalt-10 to 15 percent of rainfall.
- Weathered basalt-4 to 10 percent of rainfall.
  - * Phyllites, limestones,
  - * sandstones, quartzites,
  - * shales etc.-3 to 10 percent of rainfall.

The figures indicated above are given as a guideline and it does not automatically imply that upper limit can invariably be applied. Based upon the status of knowledge available, a value in between can be chosen.

Water Balance Approach

In this approach, all the components of water balance equation other than the rainfall recharge, are estimated using the relevant hydrological and meteorological information. The rainfall recharge is calculated by substituting these estimates in the water balance equation. A pre-requisite for successful application of this technique is very extensive and accurate hydrological and meteorological data. The water balance approach is valid for the areas where the year can be divided into monsoon and non-monsoon seasons with the bulk of rainfall occurring in former. Water balance study for monsoon and non-monsoon periods is carried out separately. The former yields an estimate of recharge coefficient and the later determines the degree of accuracy with which the components of water balance equation have been estimated.

Soil Moisture Data Based Methods

Soil moisture data based methods are the lumped and distributed model and the nuclear methods. In the lumped model, the variation of soil moisture content in the vertical direction is ignored and any effective input into the soil is assumed to increase the soil moisture content uniformly. Recharge is calculated as the remainder when losses, identified in the form of runoff and evapotranspiration, have been deducted from the precipitation with proper accounting of soil moisture deficit. In the latter model, the variation of soil moisture content in the vertical direction is accounted for and the method involves the numerical solution of partial differential equation (Richards equation) governing one-dimensional flow through unsaturated medium, with appropriate initial and boundary conditions. Nuclear techniques have been extensively used for the determination of recharge by measuring the travel of moisture through the soil column. The technique is based upon the existence of a linear relation between neutron count rate and moisture content (% by volume) for the range of moisture contents generally occurring in the unsaturated soil zone. The mixture of Beryllium (Be) and Radium (Ra) provide a convenient source of neutrons. Another method is the gamma ray transmission method based upon the attenuation of gamma rays in a medium through which it passes. The extent of attenuation or absorption is closely linked with moisture content of the soil medium. The method can be used without causing health hazards.

GEOLOGICAL TEXTURE

Texture (Geology)

Texture in geology refers to the physical appearance or character of a rock, such as grain size, shape, arrangement, and pattern at both the megascopic or microscopic surface feature level. This includes the geometric aspects and relations amongst the component particles or crystals which is called the crystallographic texture or preferred orientation. The term microstructure is preferred by some petrographers. Textures can be quantified in many ways. The most common parameter is the crystal size distribution.

Igneous Textures

Igneous textures include the rock textures occurring in igneous rocks. Igneous textures are used by geologists in determining the mode of origin igneous rocks and are used in rock classification. There are six main types of textures; phaneritic, aphanitic, porphyritic, glassy, pyroclastic and pegmatitic.

Aphanitic (*a* = not, *phaner* = visible) rocks in contrast to phaneritic rocks, typically form from lava which crystallize rapidly on or near the Earth' surface. Because extrusive rocks make contact with the atmosphere they cool quickly, so the minerals do not have time to form large crystals. The individual crystals in an aphanitic igneous rock are not distinguisable to the naked eye. Examples of aphanitic igneous rock include basalt, andesite and rhyolite.

Glassy or *vitreous* textures occur during some volcanic eruptions when the lava is quenched so rapidly that crystallization cannot occur. The result is a natural amorphous glass with few or no crystals. Examples include obsidian and pumice.

Pegmatitic texture occurs during magma cooling when some minerals may grow so large that they become massive (the size ranges from a few centimetres to several metres). This is typical of pegmatites.

Phaneritic (*phaner* = visible) textures are typical of intrusive igneous rocks, these rocks crystallized slowly below the Earth's surface. As a magma cools slowly the minerals have time to grow and form large crystals. The minerals in a phaneritic igneous rock are sufficiently large to see each individual crystal with the naked eye. Examples of phaneritic igneous rocks are gabbro, diorite and granite.

Porphyritic textures develop when conditions during cooling of a magma change relatively quickly. The earlier formed minerals will have formed slowly and remain as large crystals, whereas, sudden cooling causes the rapid crystallization of the remainder of the melt into a fine grained (aphanitic) matrix. The result is an aphanitic rock with some larger crystals (phenocrysts) imbedded within its matrix. Porphyritic texture also occurs when magma crystallizes below a volcano but is erupted before completing crystallization thus forcing the remaining lava to crystallize more rapidly with much smaller crystals.

Pyroclastic (*pyro* = igneous, *clastic* = fragment) textures occur when explosive eruptions blast the lava into the air resulting in fragmental, typically glassy material which fall as volcanic ash, lapilli and volcanic bombs.

Geologic Map Patterns for Canvas

The default installation of drawing programs such as Canvas (R) or Illustrator (R) provide insufficient patterns for designating areas on geologic maps. Although these programs make it quite simple to define a custom pattern for further use, generally this functionality is not used because of time constraints and lack of artistic talent. The USGS recognizes this and-in an attempt to standardize-provides a multitude of geologic map symbols and patterns in its Open File Report 99-430 (link to USGS publication at USGS site). The map patterns are professionally designed in Adobe Illustrator(R) format und provided by the USGS in EPS (Illustrator version) and PDF (overview) formats. Here we offer these patterns for download in Adobe Illustrator (R), Deneba Canvas (R), OpenOffice, EPS (generic version) and PDF (single fill element version) formats.

After downloading and opening the appropriate file, the map patterns become automatically available in the pattern swatch in Illustrator (R). In Canvas (R), open the file and tear away the object fill item in the tool box. Then choose the Symbol tab and expand the ink manager with the downward pointing triangle. Just drag a pattern into the square symbol definition area to make it available as a fill pattern for general use. Here is a hint from John Hunt for Freehand (R) users:

First download the appropriate PDFs from the web. Then Import the file patterns you want into Freehand (do NOT use OPEN). It then appears (be patient!) into Freehand as many individual patterns. Select the pattern tile you want. Then go to the menu Window/Inspectors/Fill which will bring a floating pallette onto your worktop. Select Tiles from the pull down menu, and hit Paste In.

Adobe Illustrator (R) version 7 or greater and Canvas (R) version 7 or greater are required. OpenOffice 1.0 can load the provided files but it does not have the notion of pattern fills. It is possible to emulate the effect by merging and intersecting shapes.

PRINCIPLES OF TERRAIN ANALYSIS

Digital terrain models have found wide applications in disciplines such as mapping, remote sensing, civil engineering, mining engineering, geology, geomorphology, military engineering, land planning, and communications. Digital Terrain Modeling: Principles and Methodology is the only authored book to provide comprehensive coverage of recent developments in the field. The volume includes an examination of data acquisition and a review of the presentation of DTMs in databases, in contour form, and in various forms of computer graphics. It explores the theories, methods, and algorithms of digital terrain modeling, and focuses on quality control and accuracy. This book discusses extensions of digital terrain models for specific problems and anticipates where this research will lead. Topics include terrain analysis, sampling strategy, acquisition methodology, surface modeling principles, triangulation algorithms, interpolation techniques, on-line and off-line quality control in data acquisition, DTM accuracy assessment and mathematical models for DTM accuracy prediction, multi-scale representation, data management, contouring, visual analysis (or visualization), the derivation of various types of terrain parameters, and future development and applications.

Terrain Analysis Considerations

Proper terrain analysis is as essential to success on the battlefield today as it was when Frederick the Great ruled Prussia over 200 years ago. It is especially relevant when a force must attack an enemy which has had time to prepare a well-organized defense. This is exactly the situation most units training at U.S. Army Combat Training Centers (CTCs) must confront. Recent CTC trends indicate that units are having difficulty conducting successful combined arms breaching operations. Staff planners can help reverse that trend by mastering categories of terrain analysis tasks which foster mobility and breach operation planning. Today's planners use modern automated tools and equipment to achieve very detailed analyses of the terrain and weather. The utility and availability of these tools continues to increase, and they provide timely support to time-strapped planners.

The Planner's Role

Most staff planners at the brigade level and below, especially the intelligence officer, assistant brigade engineer and task force engineer, realize the important role they have in conducting terrain analysis tasks in support of maneuver and breach operations planning. All BOS (Battlefield Operating Systems) representatives must consider the terrain's effects, in that the maneuver S-2 must work with Engineer, Armor, Field Artillery, Infantry, Air Defense and others to properly template and assess friendly avenues of approach, enemy obstacles and enemy battle positions (BPs). These BOS planners must survey the terrain for areas which may be used for support-by-fire (SBF) positions, attack-by-fire (ABF) positions, any covered and concealed routes into these positions, and other locations important in the planning process.

Once planners complete the initial mission analysis/IPB (Intelligence Preparation of the Battlefield), many planners think that their role as terrain analysts/integrators is complete. The reality is that the planner must perform *continuous* analysis through the entire planning, preparation and execution phases. Planners and units must be prepared to modify their plan based on this continuous evaluation of data and the situation. Planners must continually and rapidly analyze, integrate and disseminate validated enemy obstacle and dispositional intelligence. CTC trends also show a reluctance of units to modify their plans, even when indicators are present that should prompt them to make adjustments. The engineer bears considerable responsibility for this process as the force's primary expert on the terrain and its effects.

Commanders and staffs should be prepared to incorporate adjustments to the scheme of maneuver plan even during the execution of the plan, provided the recommended change makes timely, tactical sense. For instance, if R & S (Reconnaissance and Surveillance) reveals a previously unknown weakness in the enemy frontage or obstacle system, or notes a previously unidentified obstacle system on a planned route, the staff should have already built in the flexibility needed to quickly adjust the original plan.

Categories of Terrain Analysis

A TTP (tactic, technique and procedure) for grouping the terrain analysis tasks associated with planning for breaching operations places the tasks into three general categories: tasks that support template analysis, tasks that support maneuver analysis, and tasks that support planning for the breach force and assault force.

The first category, "Template Analysis" tasks, includes tasks associated with templating enemy obstacles and BPs.

The second category, "Maneuver Analysis" tasks, deals with those tasks that assist maneuver units in selecting avenues of approach (AAs) and positions which best support accomplishing their assigned tasks. It also uses information on enemy obstacles, positions, and the terrain to select the best area(s) for the point of breach (POB) and point of penetration (POP).

The final category, "Breach Analysis" tasks, involves unit-level planning for placement of vehicles and the assets used to conduct the POB and POP missions. We will examine each of these categories in more detail to emphasize the important role that planners have in influencing the decision to move and breach in the most advantageous area(s).

Terrain Task Categories

Template Analysis: The template analysis provides a doctrinal basis for estimating enemy disposition and alignment. The purpose of the tasks associated with Template Analysis is to develop an accurate situational template (SITTEMP). In creating the SITTEMP, the doctrinal template is fitted to the situation. Validated intelligence of the actual enemy composition, disposition and intent, confirmed obstacle effort, and the effects of terrain and weather help predict the most likely COA. Each separate staff planner brings a particular specialty to the table to assess and estimate their enemy counterpart's capabilities, limitations and intentions, and assists in developing several potential enemy COAs.

Planners begin by analyzing the natural and manmade features of the terrain (roads, cities, vegetation, streams, hills) and the first three military aspects of terrain (the O-C-O in OCOKA) to create a COO (Combined Obstacles Overlay). This product reflects the physical effects of the terrain and weather on the battlefield, particularly with respect to mobility. This overlay is the base product upon which the planner builds the MCOO (Modified Combined Obstacles Overlay). The planner then adds the assessments from the MCOO to the doctrinal template to create the situational template.

The engineer planner plays a key role in this process. His knowledge of obstacle tactics, the effects of terrain and weather, enemy obstacle composition, and enemy engineer capabilities assists the S-2 in accurately *estimating* enemy countermobility and survivability effort and *templating* its most likely location. Each BOS element adds similar input to the SITTEMP.

Terrain teams are especially valuable in this process, as each unit's supporting terrain team creates a detailed COO for each AO and AI. Many planners are unfamiliar with the mission, capabilities, and support provided by terrain teams, as well as how to receive or request this support. Terrain team products enable friendly forces to maximize the use of the terrain in any military operation.

Doctrinal knowledge and existing obstacle intelligence, combined with terrain products created either by the unit or a terrain team, greatly facilitate the planner's ability to select the most likely locations of enemy BPs, as well as the likely visibility and available fields of fire from those BPs. Examples of these products include weapons fan and line-of-sight (LOS) analyses.

A weapons fan product is normally created using digital map or digital image data of the area of operations plus digital elevation data. The product is a graphic depiction of the terrain that is directly visible or not visible from a position on the ground, and can, therefore, be hit with direct fire or is deadspace. The user specifies the location of the weapon, its height above the ground, the target's height above the ground, and how far out from the weapon the fan extends (the maximum effective range). The user can also draw a circle around the weapon at its maximum effective range, thus identifying where the weapon cannot reach its maximum range.

This product shows three weapons fans: two in valleys and one in the center of a broad plain. The fans are drawn with range circles at a maximum effective range of 2,500 meters. The weapons in the valleys can only fire for a short distance before their visibility ends; the center weapon can reach out to its maximum effective range in nearly every direction. Depending on the software used, the product may or may not consider the effects of vegetation on visibility. These products can be useful to any BOS. The product helps to precisely place BPs and obstacle systems for maximum advantage. For instance, the planner can position vehicles to place interlocking direct fire on an obstacle system while using terrain features to shield themselves from direct fire. A unit planner can create this product in minutes.

These products are prepared using common software (TerraBase II), and are used to determine intervisibility between two precise points on the ground.

Another example of template analysis involves knowledge of soil conditions and doctrine. The engineer planner can use detailed knowledge of the soils in the AO combined with doctrinal knowledge of enemy survivability equipment to identify areas where the enemy is likely to construct vehicle fighting positions and the total number that the enemy can realistically dig in the time available. In addition, the engineer, knowing enemy obstacle emplacement equipment and techniques, can also estimate the total amount of enemy obstacle effort available. This information, used with the previously mentioned products and the soils analysis, helps to template where and how the enemy will defend. The SITTEMP is an estimate of how the enemy will defend, and assists in driving R&S planning. As R&S confirms or denies the estimate, the S-2 and engineer refine and adjust the SITTEMP. This requires the staff to continually refine, adjust and synchronize the scheme of maneuver based on the likely enemy defensive posture.

Maneuver Analysis: The next category of terrain analysis tasks provides planning support to COA development. Maneuver analysis helps resolve questions concerning on-road and off-road trafficability, denied routes, the locations of enemy positions or obstacles, and numerous other similar considerations. Maneuver analysis helps determine the optimum locations for support-by-fire (SBF) positions, attack-by-fire (ABF) positions, the POB and POP, hide positions, and indirect fire targeting. The result of maneuver analysis is our most advantageous use of the terrain in the attack.

The terrain and how the weather affects it is a common thread binding each BOS together in the overall staff planning process. Every ground vehicle must move on the same terrain in the AO or AI. Their movement and actions are often constrained to specific AAs or mobility corridors leading to the point of breach. Planners must consider these constraints, for if the terrain or weather allows the combat power to outstrip its resupply, or if combat vehicles are forced into moving too slowly or for too long in open areas, the result can range from a slowed or incomplete mission to unacceptable losses or mission failure.

Since this set of tasks requires analysis of how vehicles will interact with the terrain, it requires a somewhat different focus. The planner can use the MCOO to classify the AAs and cross-mobility corridors as unrestricted, restricted or severely restricted. The planner can then determine the tactical viability of each AA or corridor, verifying data such as:

- Estimated vehicle/convoy speed.
- Total vehicle capacity (either in a column for a convoy or across a frontage in a maneuver formation).
- Provision for two-way traffic and resupply/rest/repair areas.
- The numbers of natural or manmade obstacles along the route.
- The numbers of gaps or water-crossing sites.
- The route's composition (gravel, dirt, pavement) and its potential for degradation in inclement weather.
- The amount of cover and concealment, where it is good and where it is not.
- The route's proximity or directness to the proposed POB and POP.
- The availability of space for the force to transition from convoy "formation" to attack formation with minimal difficulty.

The SITTEMP and MCOO (created by the engineer/S2), combined with information from unit-created products and specialized products from the terrain team, help to conduct this analysis.

The terrain team produces specific products to support this analysis, including visible area plot, elevation, slope, soil, and route maneuverability/trafficability products. The cross-country (off-road) and on-road trafficability products are especially useful in this analysis. The planner can create or use weapons fan and line-of-sight plots to identify areas the enemy could use for ambush sites and the likely locations of the ambush vehicles. The planner can use the same analysis process to identify the optimum covered/concealed routes away from the breach site toward the POP (on the far side of the objective).

Planners must conduct an integrated analysis of the weather and terrain using full situational awareness. Rainfall on the generally rocky, nonvegetated terrain at the NTC will have a different set of effects than on the generally wetter, heavily vegetated terrain in Korea or Europe. In addition, planners must separately evaluate each dry or wet gap as a potential obstacle given the composition of the force and the outlook of the weather. When there is no rainfall, maneuver units may be capable of crossing dry or wet gaps without engineer support. As rainfall increases, however, dry gaps become wet and wet gaps can become passable only with bridging assets or ford preparation, requiring bypass if such assets are unavailable. Each gap the force must cross is a drain on time and assets, and requires a significant amount of coordination. When the force must use crossing assets, staff planners must ensure there are sufficient assets to make an uninterrupted series of crossings along the planned route(s).

Breach Analysis

Breach analysis tasks are an extension of maneuver analysis tasks, and involve detailed breach planning at the unit level. The planner must evaluate the routes from the LD through both tactical and protective obstacles. The planner must consider the terrain, the weather, the SITTEMP, current obstacle intelligence, friendly movement capabilities, the optimal sites for ABF and SBF positions, friendly and enemy weapon systems capabilities, and any other information relevant to the operation. All planners must ensure they fully analyze, resource, plan and coordinate each of the breaching fundamentals (Suppress, Obscure, Secure, Reduce, Assault, or SOSR-A) (FM 3-34.2). A breach is a mission shared by many BOSs, and is critically dependent on proper evaluation and use of the terrain and weather and full participation by all BOSs.

At the breach site itself, the planner's role is critical in influencing the selection of the POB. Planners should use the MCOO, the SITTEMP, terrain team products, unit terrain products, the most recent R&S ground intelligence or the most recent imagery (from a plane, UAV or satellite) of the routes and obstacle systems that exists. These products and data allow the planner to get near-real-time data on obstacles, enemy positions, and potential breach sites.

Once the locations of the enemy obstacles and BPs are known, the planner can use the products to identify the terrain most vulnerable to direct fire from enemy BPs. The planner can use IV line, weapons fan plots and select line-of-sight analyses to choose the optimal locations for SBF and ABF positions, the most covered/concealed routes into them, and the POB. The planner can pinpoint enemy direct fire systems and obstacles, check enemy obstintel, and the POB. He can analyze the POB to determine the best assets and how to employ them in the breach drill/execution (breach force unit-level planning). The planner can use reverse LOS analysis or weapons fan plots from each proposed SBF position to identify how many BPs can observe activities at the

SBF position and along the obstacle. If the planner knows the locations of individual fighting positions, he can identify exactly which positions are within range of, and are able to observe, the obstacles and SBF positions. In this way, the planner pinpoints locations along the obstacles which may provide cover or concealment from the most direct fire weapons systems. He can select the positions that the breach force's security element can use to provide local security at the breach sites.

Planners also support the breach force commander with vegetation and soil condition overlays that indicate areas that are conducive to the breaching techniques available (mechanical, explosive, manual or electronic). Terrain products indicate the type of terrain in the area (rolling, rocky, soft) and its micro-effects on visibility, movement and breaching. The engineer and S-2 also use the MCOO to identify mobility corridors the enemy could use for a counterattack. Based on these analyses, the planner can recommend the best area(s) for the breach.

3

Satellite and Remote Sensing Processes

The term 'remote sensing' can generally be defined as "the science and art of obtaining information about an object, area, or phenomenon through the analysis of data acquired by a device not in contact with the object, area, or phenomenon under investigation."

This broad definition includes cameras, ocean buoys, and RADAR devices. These general remote sensing devices have undergone adaptations over the last century to make observations of the Earth and its physical process from aircraft and satellites.

Airplanes have been used since the 1930s to carry cameras and sensors to study the earth. Cameras collect images of part of Earth's surface, with the final aerial photograph usually consisting of a series of overlapping vertical photos that form the basis for mapping. Airplanes are also used to carry sensors.

For example, the Side-Looking Airborne Radar (SLAR) instrument is used by the United States Geological Survey (USGS) for various projects in the conterminous United States and Alaska to map geologic features, detect mineral and energy reserves, and identify potential environmental hazards.

Satellites are also used to collect images and data about the earth. Earth-observing satellites, as they are referred to, carry sensors, which are capable of recording wavelengths across the entire electromagnetic spectrum, from infrared to visible radiation. Some satellites carry sensors that collect data passively by recording radiation that is radiated or reflected from Earth's surface or atmosphere. Other satellites collect data actively by emitting radiation and then recording what is reflected back to them from Earth's surface or atmosphere.

What is Remote Sensing?

We perceive the surrounding world through our five senses. Some senses (touch and taste) require contact of our sensing organs with the objects. However, we acquire much information about our surrounding through the senses of sight and hearing which do not require close contact between the sensing organs and the external objects.

In another word, we are performing Remote Sensing all the time. Generally, Remote sensing refers to the activities of recording/observing/perceiving (sensing) objects or events at far away (remote) places.

In remote sensing, the sensors are not in direct contact with the objects or events being observed. The information needs a physical carrier to travel from the objects/events to the sensors through an intervening medium.

The electromagnetic radiation is normally used as an information carrier in remote sensing. The output of a remote sensing system is usually an image representing the scene being observed. A further step of image analysis and interpretation is required in order to extract useful information from the image. The human visual system is an example of a remote sensing system in this general sense.

In a more restricted sense, remote sensing usually refers to the technology of acquiring information about the earth's surface (land and ocean) and atmosphere using sensors onboard airborne (aircraft, balloons) or spaceborne (satellites, space shuttles) platforms.

FUNDAMENTALS OF REMOTE SENSING

The process of remote sensing involves the detection and measurement of radiation of different wavelengths reflected or emitted from distant objects or materials, by which they may be identified and categorized by class/type, substance, and spatial distribution. The background required for use of remote sensing tools may seem overwhelming at first. The decisive factor in the successful application of remote sensing data, however, need not be the technical sophistication of the user, but rather the suitability and precise use of the tool to obtain accurate and relevant data. A general grasp of the technical process that transforms electromagnetic energy into useful information can improve and expand the appropriate use of these tools. Nevertheless, depending on the application, social scientists wishing to work with remote sensing imagery would do well to partner with physical scientists with a deeper understanding of how the imagery represents physical processes on the ground.

In the broadest sense, remote sensing refers to information gathered by measuring and interpreting signals. To use perhaps the simplest analogy, the human body is constantly involved in a variety of remote sensing tasks. Hearing and vision are two obvious examples, involving the gathering and interpretation of sound and light waves, both in limited ranges of the entire sound and light spectra.

Active and *passive* remote sensing are used to describe the way sensors gather data. To illustrate the two types of sensors, consider a snapshot camera, a sensor that captures electromagnetic radiation in the visible spectrum. Outdoors in full daylight, a camera is ordinarily used as a passive sensor in that it receives reflected visible light from its surroundings and uses optics, a shutter, and film to create a lasting image. At night, on the other hand, when there is inadequate light for most cameras to capture a useable image, the camera may employ a flash. The flash is emitted from the camera and bounces off the object to illuminate it, just as an active (or radar) sensor sends a burst of energy towards its target and then receives the reflected radiation. A key factor in the choice between passive and active sensor is the relative strength of the potential signal each system must measure. For a passive system, the source of the signal is ultimately the sun, which emits electromagnetic radiation at its highest intensity between the ultraviolet and infrared ranges. In the radar wavelength ranges, however, sensors must provide a signal of sufficient intensity to travel to the earth, and return with enough strength to be distinguishable from the background "noise" from other sources.

Energy Source or Illumination (A)-A basic requirement for remote sensing is an energy source to illuminate or provides electromagnetic energy to the target of interest. For passive instruments, this is usually the sun; for so-called active instruments, the sensor itself emits an pulse of energy.

Radiation and the Atmosphere (B)-As the energy travels from its source to the target, it will come in contact with and interact with the atmosphere it passes through. This interaction also takes place a second time as the energy travels from the target to the sensor. Inevitably there is a certain degree of atmospheric scattering of radiation.

Interaction with the Target (C)-Once the energy makes its way to the target through the atmosphere, it interacts with the target depending on the properties of both the target and the radiation.

Recording of Energy by the Sensor (D)-After the energy has been scattered by, or emitted from the target, a sensor (remote-not in contact with the target) collects and records the electromagnetic radiation.

Transmission, Reception, and Processing (E)-The energy recorded by the sensor has to be transmitted, often in electronic form, to a receiving and processing station where the data are processed into an image (hardcopy or digital).

Interpretation and Analysis (F)-The processed image is interpreted, visually or digitally/electronically, to extract information about the target illuminated.

Application (G)-The final element of the remote sensing process is achieved by applying the information that has been extracted from the imagery about the target in order to better understand it, reveal some new information, or assist in solving a particular problem.

ELECTROMAGNETIC ENERGY

Remotely sensed data are collected in many regions of the electromagnetic spectrum. Data recorded from each part of the spectrum can provide distinct information on characteristics of the Earth's surface or properties of the atmosphere. For example, healthy green vegetation reflects highly in the near-infrared region of the spectrum, whereas water bodies tend to reflect only a small amount of incoming radiation in the visible region.

All remote sensing instruments collect electromagnetic radiation that is reflected, emitted, or scattered from the Earth's surface and atmosphere. So-called active sensors such as radar and lidar emit energy that bounces off the land or water surface and returns to the sensor to be recorded. The way the energy is directed or scattered by the surface, and the time it takes for the energy to return, reveals information about surface characteristics. Because of the long wave lengths employed by radar, the signals can penetrate clouds, thereby allowing scientists to record information about normally cloud-covered areas. This is an asset in tropical areas such as the Amazon River basin.

Passive sensors, on the other hand, typically rely on solar illumination of the Earth's surface, though some are equipped to detect night-time lights and gas flares. These sensors are "passive" because they do not emit their own energy, but rather rely on energy reflected or emitted from the earth's surface. Unlike radar sensors, they are unable to penetrate clouds. It is interesting to note that the visible portion of the spectrum-those wavelengths that humans can see-is a very small segment of the spectrum. Part of the strength of remote sensing is that it enables scientists to "see" portions of the spectrum that are outside the range that the human eye can detect. Scientists can combine non-visible portions with visible ones through colour composites, assigning each *band* (or portion of the spectrum detected by the instrument) the colours red, green and blue.

A sensor's bandwidth and the number and placement of bands (within the spectrum) define its *spectral properties*. Panchromatic sensors measure reflected energy in a single portion of the electromagnetic spectrum, usually the visible to near-infrared regions. Multispectral sensors, on the other hand, collect reflectance information in discrete portions of the spectrum, with each being recorded as a separate image called a band or channel. When these bands are displayed on a computer, with one band shown through each of the blue, green and red channels of the monitor, they yield a combined colour image. Landsat 7's Enhanced Thematic Mapper, for example, is a multi-spectral instrument that collects data in eight bands-three visible (one each for blue, green, and red), a near-infrared, two middle-infrared bands, a thermal-infrared and a higher spatial resolution panchromatic band. By contrast, the Moderate Imaging Spectrometer (MODIS) collects data in 36 different spectral regions, and the Airborne Visible/Infrared Imaging Spectrometer (AVIRIS) is a hyperspectral instrument that collects data in 224 spectral bands.

Ultimately, what a sensor measures is the intensity of radiation that actually reaches the sensor, which is termed the *at-satellite radiance*. Radiance values are commonly translated into digital numbers (DNs). The possible range of DNs varies between sensors, although ranges of 0-255 (for 8 bit images) and 0-1023 (for 10 bit images) are common, with higher values corresponding to greater brightness. Radiance is captured by a two dimensional array of picture elements, or *pixels*. A DN for a pixel in a specific band is determined by the intensity of the radiance captured for that particular portion of the electromagnetic spectrum. If space-based passive sensors were able to accurately, precisely and repeatedly capture the actual reflectance from a feature on the ground, regardless of the time of day, season or weather conditions, much of the hard work of image processing would be eliminated. But the reality is that the atmosphere scatters radiation that is reflected back out to space. Smoke, haze, clouds and humidity exacerbate the problem, and can block reflected energy entirely. Data from shorter wavelengths are more likely to be blocked or scattered by clouds or atmospheric particles, whereas images using sensors capturing longer wavelengths are less likely to be disturbed by atmospheric conditions between the sensor and the target object.

Platforms and Orbits

High-altitude remote sensing originated in the mid-1800s with aerial photography by balloon and, in at least one instance, the use of cameras attached to the underside of birds. Airplanes became the dominant remote sensing "platform" by the early 20th century. This practice continues to evolve for wartime, intelligence, commercial and government applications. An advantage of airborne remote sensing, is the capability of offering very high spatial resolution images (20 cm or less). The disadvantages are low coverage area and high cost per unit area of ground coverage. It is not cost-effective to map a large area using an airborne remote sensing system. Airborne remote sensing missions are often carried out as one-time operations, whereas earth observation satellites offer the possibility of continuous monitoring of the earth. The development of satellite remote sensing had greatly improved the ability to cover large areas.

There are two groups of satellites depending on the orbit in which they are placed. A *geostationary* orbit is established when a satellite is placed at a very high altitude, roughly 36,000 km above the earth's equator, and caused to orbit with the earth's rotation (called a *prograde* orbit). The altitude may vary slightly from one geostationary satellite to the next, depending upon the mass of the satellite, but, for the most part, this is a fundamental physical constraint. The rules of geometry-that is, the sight line from the satellite's position above the equator to the farthest edge of the earth's sphere-dictate that geostationary satellites can only "see" limited area at any one time. The laws of physics and capabilities of engineering limit their spatial resolution to a range of about 1 to 10 square kilometres. The total image or scene size, known as the *field of view*, is often thousands of kilometres. Thus, unless a geostationary satellite spins or turns its optics, its view is necessarily fixed. This allows for continuous monitoring, and often a very large, synoptic view of much of one entire hemisphere.

The coarse (km range) resolution versus the wide, continuous field of view constitute the main tradeoffs to consider for this orbit type. These characteristics make geostationary satellites best for collecting weather and climate data (such as cloud cover and surface temperature) and relaying communications data, although AVHRR data are used for global-and regional-scale land cover analyses.

The other group of satellites, by far the largest group of earth-orbiting satellites is with the *sun-synchronous* or *polar-orbiting*. These are lunched below the altitude of geostationary satellites closer to the earth's surface, at orbits ranging from 700km to 1000km. These satellites usually orbit at a steep inclination relative to the equator, in the direction opposite the earth's rotation, known as a *retrograde* orbit. When the satellite's orbit and earth's rotation are combined, they result in an s-shaped path relative to a map of the earth's surface. Given enough time, the orbits and rotations of the earth bring the satellite over the same location, leading to the term *exact repeat* satellites. The number of orbits between each return to the same longitude and latitude is called the *repeat cycle*.

These satellites usually orbit the earth in roughly 100 to 120 minutes, circling several times per day, returning a satellite to the same position over the earth's surface only after 2 weeks or more. The speed of motion limits the time that a satellite spends over a location, and the amount of time a scanner can "look" at any single ground cell (called the *dwell time*). Most exact repeat satellites that use passive sensors are also in sun synchronous orbits, meaning that they cross the same latitude at the same daylight time with each orbit, but with their location shifted to a different longitude. Their lower altitude allows these satellites to obtain images with spatial resolution ranging from 1-200 meters per side of a pixel, and an image width ranging from tens to thousands of kilometres per scene.

As the satellite passes over the earth's surface, its motion can be described in terms of a ground track that it follows at a certain altitude. Most satellites are *nadir looking*, meaning that their sensing equipment is aimed straight down toward the center of the earth. However, normal measurements generally include areas substantially on either side of this ground track, and that total area is called the *swath width*. Because most low-orbiting satellites follow a polar orbit, their ground tracks, and thus their swaths, are spread furthest apart at the equator, and are compressed at the poles. As a consequence, there is an overlap, called *side lap*, of neighboring swaths. Logically, this side lap is smallest at the equator and increased at the poles.

A basic understanding of ground tracks, swath widths, and side lap is helpful in designing a remote sensing experiment for a few reasons. First, weather and other temporary conditions may prevent good data acquisition for the first and/or "best" pass of a satellite directly over a given location. In this case, side lap may allow for multiple acquisitions of the same location on the surface with only a short delay, provided that the target location can be seen during two or more subsequent passes of the same satellite. This situation is more likely to occur if the sensing target is at high or low latitude than at the equator. Also, side lap may allow experimenters to obtain data for a single location in much more rapid time series than would be possible if the experimenters were to wait for the satellite to exactly repeat its path-a matter of hours, rather than weeks. Finally, side lap may allow the same location to be viewed from slightly different angles at slightly different times of the same day with neighboring orbital paths. This may provide additional information from shading caused by the sun's angle, and other factors that change with relatively small differences in position and time.

SENSORS CHARACTERISTICS

A sensor is characterized by its spectral properties (number and placement of bands), its orbital altitude and path, its swath width, and its *spatial resolution*. Spatial resolution is measured in terms of the size of one pixel projected on the ground. Spatial resolution is directly tied to the size of the features that can be resolved (or "seen") on the ground. The higher the resolution, the less likely that there will be "mixed pixels" in which radiances effectively represent an average of land cover types in the ground area represented by that pixel (e.g., half lake and half forest). Commercial high resolution sensors have a spatial resolution in the 0.6-10 meter range, medium resolution sensors fall in the 10-50 meter range, and low resolution sensors have greater than 50 meter resolution.

Until the advent of the commercial satellites IKONOS and QuickBird, with resolutions of one square meter or finer, high resolution imagery was the exclusive province of intelligence-gathering agencies. Most social science applications do not command the financial resources required to obtain such high resolution data, nor are images of this resolution generally required, except perhaps in the area of international relations, law and policy. Most social science research tends to utilize polar-orbiting satellites with medium spatial resolution, such as the Landsat, SPOT, TERRA and, more recently, AQUA satellites. They provide good spectral and ground resolution, with multiple visible, infrared, and panchromatic bands and pixel width ranging from 5 to 30 meters.

Data Processing, Interpretation and Analysis

Much of the technical work of remote sensing involves pre-processing and applying radiometric and geometric corrections to imagery to compensate for errors due to factors such as atmospheric interference of incoming radiation and sensor and data stream irregularities. Once such corrections are applied, imagery must be *georeferenced*. Georeferencing is the process of taking the image in its raw format (rows and columns of data) and linking it to the land that it covers. Images are georeferenced by linking spatially distributed control points in the satellite image to points on base maps or points referenced in the field through global positioning systems. The raster data in the image is thereby registered to a Cartesian coordinate system, and can be combined with other georeferenced data sets in a geographic information system.

The processed data can now either be visually interpreted or classified using manual or automated processes. The main elements of visual image interpretation involve gradients of tone or colour, resolution, size and shape, texture and pattern, site and association, and height and shadows. Given their knowledge of the characteristic spectral signatures of different land cover types, scientists my inspect black and white images of each band separately in order to identify features and patterns.

For many purposes, data that is collected from the earth's surface, which represents a continuous variation, needs to be categorized (de Sherbinin *et al.* 2002). Image classification is the process of creating discrete classes or categories of land cover, utilizing information from some or all of the bands to group together pixels with similar spectral signatures. Supervised classification entails providing the software with sample pixels that represent specific features, such as boreal forest, and then having the computer classify every pixel with a similar *spectral signature* as boreal forest. Analysts may also use images from different seasons in order to discriminate vegetation cover types that have different phenologies, such as deciduous and evergreen forests. In unsupervised classification, the analyst specifies the desired number of classes, and the computer automatically sorts the pixels according to their spectral signatures. The analyst then labels the resulting groups based on some local knowledge of the land cover patterns.

Once classified, it is necessary to verify or validate that the output product accurately represents the actual composition, content, structure or land surface characteristics being mapped. Validation requires either field visits, ground-truthing or comparing the classified image with existing maps or images of sufficient detail. Statistics can be derived for the classified imagery indicating the general and specific (class-wise) agreement between the pixels or classes used, letting the user know which were classified correctly and which ones were not.

Validation results are also sometimes presented as a percentage value associated with the map that communicates how accurate the map is on a per pixel basis. Since the highest confidence rankings reported by satellite land cover data sets are between 85% and 90% (for the easiest types of land cover to classify), for an image with a per pixel accuracy of 85% the probability that one pixel out of four is incorrectly classified is close to 0.50. The output of remote sensing data analysis can be presented in a variety of ways including a printout of the enhanced image itself a spatial database, summary statistics and/or graphs (Jensen 1996). The output data can be integrated with a geographic information system (GIS) database for further analysis.

REMOTE SENSING SATELLITES

United States Government Remote Sensing Programs

The United States began the current phase of Earth observation from space with the launch of the first Landsat satellite (ERTS-1/Landsat-1) in 1972. Currently, the United States has ten Earth observing satellites in orbit. Three of these are NOAA satellites (NOAA-J, NOAA-K, and NOAA-L) that comprise the NOAA Polar Operational Environmental System (POES). These NOAA satellites all carry the Advanced Very High Resolution Radiometer (AVHRR) sensor that is used for measuring vegetation densities, crop yields, ocean temperatures, forest fire danger zones, and snow cover.

The other seven are NASA satellites: Landsat 5, Landsat 7, Terra, Tropical Rainfall Measuring Mission (TRMM), Earth Probe—TOMS, Quick Scatterometer (QuikScat) and Earth Observing-1 (EO-1). Landsat 7 is used for general Earth observations including forestry, crop monitoring, land cover, land use, and watersheds.

It carries the Enhanced Thematic Mapper Plus (ETM+), which boasts improved data collection capabilities from previous Landsat missions. Terra differs from Landsat in that it is dedicated to observing process more than land features. Terra carries five different sensors, each having unique applications, ranging from land temperature and snow/glacier cover measurements (ASTER), to cloud cover and radiant energy (CERES), to pollution measurements (MOPITT), to aerosol and smoke plume imaging (MISR), to ocean productivity and temperature ranges (MODIS). The five sensors carried on the TRMM satellite are all committed to record tropical and subtropical atmospheric parameters such as rainfall, lightning, and cloud cover. The TOMS sensor carried on the Earth Probe craft observes rates of ozone depletion, daily UV exposure, UV-absorbing aerosols and data on dust, smoke, and ash in the troposphere. The SeaWinds sensor carried on QuikScat uses specialized radar to measure near-surface wind speed and direction.

International Remote Sensing Programs

International efforts have pioneered the development of active remote sensing satellites. Canadian Space Agency's RADARSAT-1 and the European Space Agency's Remote Sensing satellites (ERS-1 and-2) carry radar sensors that emit and record microwave signals, permitting observations independent of weather or daylight conditions.

France, India, Russia, Japan, and the China-Brazil team all operate successful passive satellite programs. France controls the Systeme Pour l'Observation de la Terre (SPOT), which is comprised of satellites SPOT 1, SPOT 2, and SPOT 4. The payload of the SPOT satellites consists of two high-resolution-visible (HRVIR) sensors that can operate in either panchromatic (SPOT pan) or multispectral (SPOT xs) modes with a resolution of 10-20 meters depending on the mode. SPOT has many applications, including land use, water resources research, coastal monitoring, crop production, and deforestation. SPOT-4 also carries a wide-angle (2000 km) system referred to as VEGETATION that will be used for international crop monitoring. The Indian Space Research Organization (ISRO) currently operates four Earth-observing satellites; the most recently launched (IRS-P4/Ocenasat) focuses on oceanic research. Other Earth-observing systems (EOS) include Russia's Resurs-O1 series, Japan's ADEOS system, and the CBERS satellite that is operated jointly by China and Brazil.

Commercial Satellite Systems

The U.S. government has encouraged the development of independent commercial satellites and many U.S. companies have designed and launched their own satellites. Orbital Imaging Corporation (ORBIMAGE) and Space Imaging, Inc. both have successful satellites in orbit that carry high-resolution sensors. ORBIMAGE operates two satellites. The first, OrbView-1, is designed to monitor atmosphere. The second, OrbView-2 (SeaStar), carries a sensor called SeaWiFS (Sea-viewing Wide Field-of-view Sensor) that was developed in conjunction with NASA. SeaWiFS is designed to monitor ocean temperature and productivity. Space Imaging operates one satellite, IKONOS, which boasts 1-meter resolution capabilities in the panchromatic (black and white) range and 4-meter resolution in the multispectral range. IKONOS has applications ranging from imaging coral reefs to aiding highway planning.

Remote Sensing Applications

The potential applications of these satellite sensors are vast. This section briefly describes some of the possible environmental applications, focusing on environmental enforcement, land use planning, forestry, agriculture, water resources, fisheries, wetlands, watersheds, climate change, and disaster management.

Environmental Enforcement

The U.S. Environmental Protection Agency (EPA) conducts four types of satellite and aerial remote sensing projects to support the Comprehensive Environmental Response, Compensation, and Liability Act (CERCLA, also known as the Superfund Act), the Resource Conservation and Recovery Act

(RCRA), and in other EPA regulatory programs and investigations. The projects are: (1) emergency response to hazardous material release that requires rapid site assessment; (2) single-date analysis to update old data on the current conditions of the site; (3) intensive site analysis of current and historic images, to obtain an understanding of changing conditions over time; and (4) waste site inventories over large areas to locate possible disposal sites. Images from these projects can standalone or be used in conjunction with topographic maps, digital elevation data, and other features stored in GIS databases.

Further use of remote sensing (both satellite and aerial photography) as a tool in environmental forensics is discussed in a two-part paper by Brilis, et al. The paper outlines the general approach to be followed when planning the use of remote sensing in environmental forensics. The accuracy of locational data and the use of metadata are identified as two critical items to ensure that a final image can withstand veracity issues when used for courtroom presentation.

Recently, interest has developed in using satellites to monitor and enforce multilateral environmental agreements (MEAs), such as the Kyoto Protocol. Remote sensing data may be used in the future to ensure compliance with MEA requirements by both direct enforcement and by more indirect means, such as deterring non compliance through high levels of transparency.

Land Use Planning and Change

Passive sensors, including those on the NOAA-AVHRR, IKONOS, Landsat, and SPOT satellites, are used in a broad range of forest and land use applications. These applications include estimations of primary production, biomass, crop yields, and to chart vegetation type, deforestation, desertification, forest boundaries, forest harvest, soil erosion, and bush or forest fires. Landsat 7's EMT+ sensor is especially useful in studying land use change because its data has been archived since the first Landsat mission in 1972. Passive sensors have also been used to observe and monitor changes associated with storm, flood, and fire damage.

Forestry

Forestry applications for passive remote sensors include tree species surveys, monitoring clear cut operations, planning and observing burn areas, and studying successional forest growth. The U.S. Forest Service (USFS) relies primarily on the data from Landsats 5 & 7 for forest monitoring because of the low cost and large scene size. Landsat data is particularly applicable to forest change monitoring because data from previous Landsat missions is archived and available for accurate comparison with data from the current Landsat mission.

The USFS also uses SPOT data in conjunction with Landsat data to increase the level of detail in sensitive areas. Active sensors, carried on the RADARSAT and ERS satellites, are capable of making course scale distinctions between cover types such as late successional forests, newly planted forests, clear cut forests, burn areas, agricultural areas, and deserts. Active sensors are valuable tools for monitoring crop regulation compliance, forest clearing, and for taking general inventories of world forest densities.

Agriculture

The United States Department of Agriculture (USDA) is conducting research to determine the potential uses of remote sensing (both aerial and satellite) in the agricultural sector. Promising applications include measuring leaf area indices (LAI—a quantitative indicator of leaf stress), identifying soil properties by their spectral signals, evaluating crop productivity, and providing a valuable data source for crop simulation models. A high-tech type of farming known as "precision agriculture," uses satellite data to characterize specific sections of a field by certain variables (such as water or nutrient levels). Once the characteristics and geographic coordinates of the field section are in a computer, additions such as water, pesticides, and fertilizers can be efficiently controlled in response to the specific needs of each section thereby reducing the amount of pollutants introduced to the environment while producing healthier crops.

Water Resources and Fisheries

SeaWiFS is designed to monitor oceans and track water indicators such as turbidity, sediment load and transport, primary production by marine phytoplankton, algal blooms, chlorophyll content, dissolved oxygen, and pH. Other applications include managing coral reefs, monitoring pollution and oil spills, and characterizing and monitoring short-term and long-term fish habitat. Terra's MODIS and AVHRR sensors record observations of sea surface temperature, which is directly relevant to fisheries due to individual species' temperature requirements for survival and propagation. The sensor may also help predict migration routes. Active sensing technologies are capable of measuring sea level, wave height, surface wind speed, current fronts, eddies, and surface temperature, as well as locating ocean floor features such as trenches and seamounts. Active sensors have also been used to track oil spills, effluent discharges, and algal blooms.

Wetlands and Watersheds

Wetlands monitoring may employ a combination of land-observation and ocean-observation satellites. ETM+ data can be used to delineate wetland areas, make topographical observations, and to detect illegal development. Active systems can provide consistent and accurate observations of dynamic wetland parameters such as tidal and seasonal patterns, climate, hydrology, topography, vegetation, and soil type. Satellite data and images can also be used to delineate the flow of water through watersheds, and can even be used to track pollutants. Furthermore, using algal productivity as an indicator, scientists are able to monitor whether high levels of nutrients pollute areas of a watershed.

Climate Change

In the past decade, various ozone-monitoring sensors have been launched to study global climate cycles. These include the TOMS sensor and many of the sensors on Terra and future EOS satellites. AVHRR data from NOAA's POES satellites is used in conjunction with RADARSAT to monitor the polar ice sheets and iceberg movements. The EOS satellites, beginning with the Terra, were designed specifically for monitoring climate conditions, including the observation of aerosols, cloud cover, fires, ocean productivity, pollution, solar radiation, sea ice, and snow cover.

Disaster Management and Emergency Response

Remote sensing technologies can provide the government with the ability to avoid much of the damage caused by unforeseen natural disasters. While weather satellites have monitored hurricanes and tornados since the 1960s, other satellite sensors, such as ETM+ and MODIS, have potential applications for disaster management and response. Scientists have used ETM+ data to monitor patterns in floods, droughts, beach erosion, and volcanic activity over time. MODIS and ASTER data can forecast severe weather with a great degree of reliability, potentially saving states millions of dollars in unnecessary evacuation and emergency response. For forest fire emergencies, TOMS data can identify and monitor the occurrence of forest fires, especially in remote areas, while AVHRR data can create maps denoting fire-susceptible areas. NOAA-POES and NOAA-GOES (Geostationary Operational Environmental Satellite) are used to make weather observations including predicting local weather, tracking weather in real time globally and locally, understanding and predicting hurricanes and other severe weather, studying phenomena such as El Nino, La Nina, the Gulf Stream and other global current patterns, and observing the dynamics between the land temperature, ocean processes, and the atmosphere.

SATELLITE REMOTE SENSING

In this CD, you will see many remote sensing images around Asia acquired by earth observation satellites. These remote sensing satellites are equipped with sensors looking down to the earth. They are the "eyes in the sky" constantly observing the earth as they go round in predictable orbits.

Effects of Atmosphere

In satellite remote sensing of the earth, the sensors are looking through a layer of atmosphere separating the sensors from the Earth's surface being observed. Hence, it is essential to understand the effects of atmosphere on the electromagnetic radiation travelling from the Earth to the sensor through the atmosphere. The atmospheric constituents cause wavelength dependent absorption and scattering of radiation. These effects degrade the quality of images. Some of the atmospheric effects can be corrected before the images are subjected to further analysis and interpretation.

A consequence of atmospheric absorption is that certain wavelength bands in the electromagnetic spectrum are strongly absorbed and effectively blocked by the atmosphere. The wavelength regions in the electromagnetic spectrum usable for remote sensing are determined by their ability to penetrate atmosphere. These regions are known as the atmospheric transmission windows. Remote sensing systems are often designed to operate within one or more of the atmospheric windows.

These windows exist in the microwave region, some wavelength bands in the infrared, the entire visible region and part of the near ultraviolet regions. Although the atmosphere is practically transparent to x-rays and gamma rays, these radiations are not normally used in remote sensing of the earth.

Optical and Infrared Remote Sensing

In Optical Remote Sensing, optical sensors detect solar radiation reflected or scattered from the earth, forming images resembling photographs taken by a camera high up in space. The wavelength region usually extends from the visible and near infrared (commonly abbreviated as VNIR) to the short-wave infrared (SWIR). Different materials such as water, soil, vegetation, buildings and roads reflect visible and infrared light in different ways. They have different colours and brightness when seen under the sun. The interpretation of optical images require the knowledge of the spectral reflectance signatures of the various materials (natural or man-made) covering the surface of the earth. There are also infrared sensors measuring the thermal infrared radiation emitted from the earth, from which the land or sea surface temperature can be derived.

Microwave Remote Sensing

There are some remote sensing satellites which carry passive or active microwave sensors. The active sensors emit pulses of microwave radiation to illuminate the areas to be imaged. Images of the earth surface are formed by measuring the microwave energy scattered by the ground or sea back to the sensors. These satellites carry their own "flashlight" emitting microwaves to illuminate their targets. The images can thus be acquired day and night. Microwaves have an additional advantage as they can penetrate clouds. Images can be acquired even when there are clouds covering the earth surface.

A microwave imaging system which can produce high resolution image of the Earth is the synthetic aperture radar (SAR). The intensity in a SAR image depends on the amount of microwave backscattered by the target and received by the SAR antenna. Since the physical mechanisms responsible for this backscatter is different for microwave, compared to visible/infrared radiation, the interpretation of SAR images requires the knowledge of how microwaves interact with the targets.

REMOTE SENSING IMAGES

Remote sensing images are normally in the form of digital images. In order to extract useful information from the images, image processing techniques may be employed to enhance the image to help visual interpretation, and to correct or restore the image if the image has been subjected to geometric distortion, blurring or degradation by other factors.

There are many image analysis techniques available and the methods used depend on the requirements of the specific problem concerned. In many cases, image segmentation and classification algorithms are used to delineate different areas in an image into thematic classes. The resulting product is a thematic map of the study area. This thematic map can be combined with other databases of the test area for further analysis and utilization.

Visual System

The human visual system is an example of a remote sensing system in the general sense. The sensors in this example are the two types of photosensitive cells, known as the cones and the rods, at the retina of the eyes. The cones are responsible for colour vision. There are three types of cones, each being sensitive to one of the red, green, and blue regions of the visible spectrum. Thus, it is not coincidental that the modern computer display monitors make use of the same three primary colours to generate a multitude of colours for displaying colour images. The cones are insensitive under low light illumination condition, when their jobs are taken over by the rods. The rods are sensitive only to the total light intensity. Hence, everything appears in shades of grey when there is insufficient light.

As the objects/events being observed are located far away from the eyes, the information needs a carrier to travel from the object to the eyes. In this case, the information carrier is the visible light, a part of the electromagnetic spectrum. The objects reflect/scatter the ambient light falling onto them. Part of the scattered light is intercepted by the eyes, forming an image on the retina after passing through the optical system of the eyes. The signals generated at the retina are carried via the nerve fibres to the brain, the central processing unit (CPU) of the visual system. These signals are processed and interpreted at the brain, with the aid of previous experiences.

When operating in this mode, the visual system is an example of a "Passive Remote Sensing" system which depends on an external source of energy to operate. We all know that this system won't work in darkness. However, we can still see at night if we provide our own source of illumination by carrying a flashlight and shining the beam towards the object we want to observe. In this case, we are performing "Active Remote Sensing", by supplying our own source of energy for illuminating the objects.

The Planet Earth

The planet Earth is the third planet in the solar system located at a mean distance of about 1.50×10^8 km from the sun, with a mass of 5.97×10^{24} kg. Descriptions of the shape of the earth have evolved from the flat-earth model, spherical model to the currently accepted ellipsoidal model derived from accurate ground surveying and satellite measurements. A number of reference ellipsoids have been defined for use in identifying the three dimensional coordinates (i.e. position in space) of a point on or above the earth surface for the purpose of surveying, mapping and navigation. The reference ellipsoid in the World Geodetic System 1984 (WGS-84) commonly used in satellite Global Positioning System (GPS) has the following parameters:

- Equatorial Radius = 6378.1370 km
- Polar Radius = 6356.7523 km

The earth's crust is the outermost layer of the earth's land surface. About 29.1% of the earth's crust area is above sea level. The rest is covered by water. A layer of gaseous atmosphere envelopes the earth's surface.

The Earth's Atmosphere

The earth's surface is covered by a layer of atmosphere consisting of a mixture of gases and other solid and liquid particles. The gaseous materials extend to several hundred kilometres in altitude, though there is no well defined boundary for the upper limit of the atmosphere. The first 80 km of the atmosphere contains more than 99% of the total mass of the earth's atmosphere.

Vertical Structure of the Atmosphere

The vertical profile of the atmosphere is divided into four layers: troposphere, stratosphere, mesosphere and thermosphere. The tops of these layers are known as the tropopause, stratopause, mesopause and thermopause, respectively.

Troposphere: This layer is characterized by a decrease in temperature with respect to height, at a rate of about 6.5ºC per kilometer, up to a height of about 10 km. All the weather activities (water vapour, clouds, precipitation) are confined to this layer. A layer of aerosol particles normally exists near to the earth surface. The aerosol concentration decreases nearly exponentially with height, with a characteristic height of about 2 km.

Stratosphere: The temperature at the lower 20 km of the stratosphere is approximately constant, after which the temperature increases with height, up to an altitude of about 50 km. Ozone exists mainly at the stratopause. The troposphere and the stratosphere together account for more than 99% of the total mass of the atmosphere.

Mesosphere: The temperature decreases in this layer from an altitude of about 50 km to 85 km.

Thermosphere: This layer extends from about 85 km upward to several hundred kilometres. The temperature may range from 500 K to 2000 K. The gases exist mainly in the form of thin plasma, i.e. they are ionized due to bombardment by solar ultraviolet radiation and energetic cosmic rays.

The term upper atmosphere usually refers to the region of the atmosphere above the troposphere. Many remote sensing satellites follow the near polar sun-synchronous orbits at a height around 800 km, which is well above the thermopause.

Atmospheric Constituents

The atmosphere consists of the following components:

Permanent Gases: They are gases present in nearly constant concentration, with little spatial variation. About 78% by volume of the atmosphere is nitrogen while the life-sustaining oxygen occupies 21%. The remaining one percent consists of the inert gases, carbon dioxide and other gases.

Gases with Variable Concentration: The concentration of these gases may vary greatly over space and time. They consist of water vapour, ozone, nitrogeneous and sulphurous compounds.

Solid and liquid particulates: Other than the gases, the atmosphere also contains solid and liquid particles such as aerosols, water droplets and ice crystals. These particles may congregate to form clouds and haze.

ELECTROMAGNETIC RADIATION

Electromagnetic waves are energy transported through space in the form of periodic disturbances of electric and magnetic fields.

All electromagnetic waves travel through space at the same speed, $c = 2.99792458 \times 10^8$ m/s, commonly known as the speed of light. An electromagnetic wave is characterized by a frequency and a wavelength. These two quantities are related to the speed of light by the equation,

speed of light = frequency × wavelength

The frequency (and hence, the wavelength) of an electromagnetic wave depends on its source.

There is a wide range of frequency encountered in our physical world, ranging from the low frequency of the electric waves generated by the power transmission lines to the very high frequency of the gamma rays originating from the atomic nuclei. This wide frequency range of electromagnetic waves constitute the Electromagnetic Spectrum.

The Electromagnetic Spectrum

The electromagnetic spectrum can be divided into several wavelength (frequency) regions, among which only a narrow band from about 400 to 700 nm is visible to the human eyes. Note that there is no sharp boundary between these regions.

Wavelength units: 1 mm = 1000 μm; 1 μm = 1000 nm.

- Radio Waves: 10 cm to 10 km wavelength.

- Microwaves: 1 mm to 1 m wavelength. The microwaves are further divided into different frequency (wavelength) bands: (1 GHz = 10^9 Hz)
  - o P band: 0.3-1 GHz (30-100 cm)
  - o L band: 1-2 GHz (15-30 cm)
  - o S band: 2-4 GHz (7.5-15 cm)
  - o C band: 4-8 GHz (3.8-7.5 cm)
  - o X band: 8-12.5 GHz (2.4-3.8 cm)
  - o Ku band: 12.5-18 GHz (1.7-2.4 cm)
  - o K band: 18-26.5 GHz (1.1-1.7 cm)
  - o Ka band: 26.5-40 GHz (0.75-1.1 cm)
- Infrared: 0.7 to 300 μm wavelength. This region is further divided into the following bands:
  - o Near Infrared (NIR): 0.7 to 1.5 μm.
  - o Short Wavelength Infrared (SWIR): 1.5 to 3 μm.
  - o Mid Wavelength Infrared (MWIR): 3 to 8 μm.
  - o Long Wanelength Infrared (LWIR): 8 to 15 μm.
  - o Far Infrared (FIR): longer than 15 μm.

The NIR and SWIR are also known as the Reflected Infrared, referring to the main infrared component of the solar radiation reflected from the earth's surface. The MWIR and LWIR are the Thermal Infrared.

- Visible Light: This narrow band of electromagnetic radiation extends from about 400 nm (violet) to about 700 nm (red). The various colour components of the visible spectrum fall roughly within the following wavelength regions:
  - o Red: 610-700 nm
  - o Orange: 590-610 nm
  - o Yellow: 570-590 nm
  - o Green: 500-570 nm
  - o Blue: 450-500 nm
  - o Indigo: 430-450 nm
  - o Violet: 400-430 nm
- Ultraviolet: 3 to 400 nm
- X-Rays and Gamma Rays.

Photons

According to quantum physics, the energy of an electromagnetic wave is quantized, i.e. it can only exist in discrete amount. The basic unit of energy for an electromagnetic wave is called a photon. The energy E of a photon is proportional to the wave frequency f,

$E = h f$

where the constant of proportionality h is the Planck's Constant,

$h = 6.626 \times 10^{-34}$ J s.

Effects of Atmosphere

When electromagnetic radiation travels through the atmosphere, it may be absorbed or scattered by the constituent particles of the atmosphere. Molecular absorption converts the radiation energy into excitation energy of the molecules. Scattering redistributes the energy of the incident beam to all directions. The overall effect is the removal of energy from the incident radiation. The various effects of absorption and scattering are outlined in the following sections.

Atmospheric Transmission Windows

Each type of molecule has its own set of absorption bands in various parts of the electromagnetic spectrum. As a result, only the wavelength regions outside the main absorption bands of the atmospheric gases can be used for remote sensing. These regions are known as the Atmospheric Transmission Windows.

The wavelength bands used in remote sensing systems are usually designed to fall within these windows to minimize the atmospheric absorption effects. These windows are found in the visible, near-infrared, certain bands in thermal infrared and the microwave regions.

Effects of Atmospheric Absorption

Atmospheric absorption affects mainly the visible and infrared bands. Optical remote sensing depends on solar radiation as the source of illumination. Absorption reduces the solar radiance within the absorption bands of the atmospheric gases. The reflected radiance is also attenuated after passing through the atmosphere. This attenuation is wavelength dependent. Hence, atmospheric absorption will alter the apparent spectral signature of the target being observed.

Effects of Atmospheric Scattering

Atmospheric scatterring is important only in the visible and near infrared regions. Scattering of radiation by the constituent gases and aerosols in the atmosphere causes degradation of the remotely sensed images. Most noticeably, the solar radiation scattered by the atmosphere towards the sensor without first reaching the ground produces a hazy appearance of the image. This effect is particularly severe in the blue end of the visible spectrum due to the stronger Rayleigh Scattering for shorter wavelength radiation.

Furthermore, the light from a target outside the field of view of the sensor may be scattered into the field of view of the sensor. This effect is known as the adjacency effect. Near to the boundary between two regions of different brightness, the adjacency effect results in an increase in the apparent brightness of the darker region while the apparent brightness of the brighter region is reduced. Scattering also produces blurring of the targets in remotely sensed images due to spreading of the reflected radiation by scattering, resulting in a reduced resolution image.

Airborne Remote Sensing

In airborne remote sensing, downward or sideward looking sensors are mounted on an aircraft to obtain images of the earth's surface. An advantage of airborne remote sensing, compared to satellite remote sensing, is the capability of offering very high spatial resolution images (20 cm or less). The disadvantages are low coverage area and high cost per unit area of ground coverage. It is not cost-effective to map a large area using an airborne remote sensing system. Airborne remote sensing missions are often carried out as one-time operations, whereas earth observation satellites offer the possibility of continuous monitoring of the earth.

Analog aerial photography, videography, and digital photography are commonly used in airborne remote sensing. Synthetic Aperture Radar imaging is also carried out on airborne platforms.

Analog photography is capable of providing high spatial resolution. The interpretation of analog aerial photographs is usually done visually by experienced analysts. The photographs may be digitized using a scanning device for computer-assisted analysis.

Digital photography permits real-time transmission of the remotely sensed data to a ground station for immediate analysis. The digital images can be analysed and interpreted with the aid of a computer.

In spaceborne remote sensing, sensors are mounted on-board a spacecraft (space shuttle or satellite) orbiting the earth. At present, there are several remote sensing satellites providing imagery for research and operational applications. Spaceborne remote sensing provides the following advantages:

- Large area coverage;

- Frequent and repetitive coverage of an area of interest;
- Quantitative measurement of ground features using radiometrically calibrated sensors;
- Semiautomated computerised processing and analysis;
- Relatively lower cost per unit area of coverage.

Satellite imagery has a generally lower resolution compared to aerial photography. However, very high resolution imagery (up to 1-m resolution) is now commercially available to civilian users with the successful launch of the IKONOS-2 satellite in September 24, 1999.

Satellite Orbits

A satellite follows a generally elliptical orbit around the earth. The time taken to complete one revolution of the orbit is called the orbital period. The satellite traces out a path on the earth surface, called its ground track, as it moves across the sky. As the earth below is rotating, the satellite traces out a different path on the ground in each subsequent cycle. Remote sensing satellites are often launched into special orbits such that the satellite repeats its path after a fixed time interval. This time interval is called the repeat cycle of the satellite.

COMMERCIAL REMOTE SENSING

The commercial satellite remote sensing market was initiated in 1972, around the time that the launch of the first Landsat mission (then referred to as ERTS-1) was being discussed. The success of commercial weather and communications satellites led the U. S. to believe that a land-observing satellite would eventually be able to pay for itself as private markets for the data grew. While the commercialization of Landsat has not been viewed as a success, it paved the way for the growth of the data distribution and value-added product industry sectors.

Private companies launching satellites can profit both as data collectors, and as intermediaries between raw data providers and the ultimate consumers. However, few private companies have launched successful Earth observation satellites.

Between 1993 and 2000, following the passage of the 1992 Land Remote Sensing Commercialization Act, NOAA issued only seventeen licenses for private commercial satellites. Of the first four companies to launch private satellites, the two successes have been Space Imaging, Inc. and Orbital Imaging Corp. (Orb Image).

Associated Geospatial Technologies

Image-based GIS and photogrammetry comprised 69% of the geospatial activities market in 2000, with mapping, civil government, environmental, transportation, and national/global security markets controlling the highest percentage of sales. The market leader, Environmental Systems Research Institute (ESRI), controlled nearly $300 million of the $845 million total worldwide GIS software market in 1999. ESRI products provide a broad variety of applications to industries ranging from telecommunications and engineering to humanitarian assistance and environmental conservation. Intergraph holds the second largest market share, $238.7 million, of the total industry software market. Intergraph is the industry leader in providing GIS services and products to the utilities and telecommunications sectors, and is also an industry leader in public safety, transportation, and mapping sectors.

Due to the significance of the GIS market, commercial satellite companies such as Space Imaging and OrbImage have formed business relationships with GIS providers, predominantly ESRI. These two growing industries have both been aided by advances in the integration of remote sensing imagery and geographic information systems. Satellite data and images provide geospatially-referenced data for inclusion into GIS layers and can be used to create digital elevation models or other applications for GIS. ESRI's software can ingest, enhance, and classify IKONOS imagery and utilize it just like any other data layer in a GIS analysis. The imagery can serve as an incredibly detailed basemap upon which other layers are laid, or it can be used as an up-to-date data source from which various land cover and elevation features are extracted to populate multiple GIS layers.

The ASPRS/NASA Ten-Year Industry Forecast Currently, NASA and the American Society for Photogrammetry & Remote Sensing (ASPRS) are conducting a 10-year market survey of the remote sensing industry as defined by the "Space Act Agreement" between NASA and ASPRS. In the first phase of the study, the team determined the baseline forecast of the U.S. Remote Sensing Industry (RSI) and associated geospatial activities. Four key findings emerged from the first phase of the study:

(1) The U.S. Remote Sensing Industry in 2000 was valued at $2.2 billion and is expected to grow at an average of 10-15% per year over the next five years, 2010;

(2) Currently the photogrammetry market is the largest Geospatial Market in terms of sales. Research and development for remote sensing, however, is considerably larger than that for photogrammetry or image-based GIS;

(3) Across all three sectors, the most active commercial markets are mapping/geography, environment, civil government, national/global security, and transportation. Environmental applications were rated one of the top four applications in all three sectors; and

(4) For the government sector, mapping, earth natural science research, and natural resource management were found to be the three most important missions.

Obstacles to Industry Growth

Numerous obstacles block the full realization of the remote sensing market. The Carnegie Endowment for International Peace identifies four critical factors that will ultimately decide the size of the market: the extent of government interference, the cost of commercial imagery, the time the data takes to reach the consumer, and the ability of the market to educate and interest consumers. Workshops have further addressed the problems related to acceptance of remote sensing technologies. Issues ranging from a lack of access to standardizing data sets to "disconnects" between data providers and data users have been recognized, but not yet solved. However, one of the most pivotal obstacles, which must be overcome before others can be addressed, is that of acceptance of satellite data in a courtroom.

GENERAL IDEA ABOUT ELECTROMAGNETIC SPECTRUM

The idea of the electromagnetic spectrum is one of those great scientific syntheses of observable natural phenomena that should be part of the knowledge of any scientifically educated person. Other similar syntheses are things like the kinetic-molecular theory of matter, the concept of the vastness of the Universe and our relation to it, the idea of deep time and the evolutionary development of life, the DNA code of life and the ideas of basic biochemistry.

Since the information technologist makes use of the electromagnetic spectrum a great deal, it is appropriate to devote a lecture to it in our course.

Electricity and Magnetism

These were two disparate ideas that were first discovered in antiquity: the Greeks gave the name (as in so many areas) through their term for amber. Amber is a naturally occurring fossizilized tree sap, made famous in the movie Jurassic Park. Anyway, the Greeks had observed that amber when rubbed could attract small pieces of dust or other light trash. Also, static electricity sparks were also known.

Naturally occurring magnetic material ("lodestone" or a type of iron oxide with a higher concentration of iron than usual) was also known to the ancients. It is said that the Chinese were aware that a piece of lodestone could be floated on a piece of cork, and would stabilize to a consistent North-South orientation, thus making possible a compass.

Hewever, it wasn't until the 1700s that the beginnings of our modern "scientific" approach to electricity and magnetism began.

Alessandro Volta noted a connection between electricity and life when he made dissected frogs legs twitch by touching them with dissimilar metals. [This inspired Mary Shelley, the wife of the famed poet Percy Byssche Shelley, to write a famous novel, whose theme was "There are things that Man is not meant to know..."] Anyway, Volta had been experimenting with placing dissimilar metals, *e.g.* copper and zinc, into salt solutions, and obtaining an electric current.

It was early shown that if wire was wrapped into a coil, and connected to a battery, that a magnet was produced in the area close to the coil. If a piece of iron was placed inside the coil, the field was greatly magnified, giving a device called an electromagnet. If the iron was a hard steel alloy, it would retain the magnetism after the current was removed, allowing the manufacture of permanent magnets. Lodestones were quite weak compared to artificial magnets.

Hans Christian Oersted observed that if a wire was wrapped around a compass, the compass would deflect.

This was a unique phenomenon. The uniqueness came from the fact that the interaction was at right angles. The compass needle didn't line itself up with the wire, but at right angles to it. This was the first "vector cross product" relationship discovered. All other interactions hitherto known acted in a straight line between centers: gravitational attraction, electric force and magnetism.

Scientists had been experimenting with pure electric force also. To observe pure electric force out in the open requires charges of thousands of volts, which, fortunately, are easily obtained by friction. When you give someone a shock by "static electricity" you are using thousands of volts. It is not lethal because the amount of electric current is small.

[I use the term "scientists" here. In those days, in the 18th and 19th centuries, the usual term was "philosophers" or "natural philosophers". Our "science" was their "natural philosophy."]

[One can make water analogies of these ideas; in fact any electrical device can have a water analogy. One can think of electric charge as being a quantity of water, current as the rate of flow of water, and voltage as the force or pressure propelling the water. The force or pressure comes from a source of energy; either the water must be pumped up to a height to give it pressure-that is, the water has potential energy, or else a pump must take energy from some other source to force the water to move directly. Our city water system uses a combination of both methods. Strictly speaking, voltage is called potential difference, and represents the potential energy per unit charge, measured between two points in space.]

These studies were being carried out mainly in Europe, especially France and Italy. England was mainly preoccupied with the developing industrial revolution, and its scientific workers were clarifying ideas about energy and thermodynamics, motivated by the development of the steam engine.

America was regarded by European intellectuals as a land of wild Indians and stagecoaches, if they thought about it at all. The idea that someone in America could do Natural Philosophy seemed absurd. However, Benjamin Franklin in the middle of the 1700s had a brief period in his life when he had made enough money from his printing shop to semi retire, but hadn't become embroiled in the American Revolution as representative in France yet.

He experimented with electricity, leading up to his famous kite experiment in which he showed that lightning was the same basic thing as static electricity created in the lab. Since he knew that lab static dissipated quickly if the charged object had sharp points on it, he thought that maybe lightning could be dissipated before the actual stroke if sharp objects electrically connected to the earth were to be placed on high roofs. This was shown to be effective. Some feel that this vivid demonstration of the ability of man to evidently interfere with the "will of God" successfully was a significant force leading to the growth of secularism in Christendom. There was, after all, no more vivid demonstration of the "will of God" than to be struck down by a lightning bolt!

A similar formula was shown to work for idealized magnetic poles:

The constant mu zero is used to make the numbers work out. M1 and M2 are the magnetic pole strengths. [In fact this equation is an idealization more so than the electric equation above. An isolated magnetic pole is an unachievable abstraction-in practice magnets always exist as dipoles-N and S poles fairly close together. Electric poles can in principle be separated as much as needed.] By the middle 1800s quite a bit of electrical technology had been worked out. Things like telegraph lines were beginning to appear, allowing instant communication over vast distances. In fact, the mid 19th century saw communication finally surpass that under the Roman Empire 2000 years earlier!

Electric batteries, motors, generators existed. Electric power stations were starting to appear.

James Clerk Maxwell studied all the disparate physics and engineering involved in this field, and formulated a theoretical formulation that tied it all togther. This led to the great triumph of "field theory" in which each point in space could be thought of as posessing a magnetic and electric force; that is, there existed an electric and a magnetic field.

Oersted had shown that these were related, and the electric motor and generator were the practical embodiment of this discovery. Physics students learn about these when they study electromagnetism in 2nd or 3rd year. A good part of the early technical education of a would-be physicist is to learn the necessary mathematical tools, that is, integral and differential calculus, to enable manipulation of these equations. A physicist has to learn how to apply these to specific real-world situations (albeit a bit idealized for pencil and paper calculations). This theory was a revolution in the 1870s when it appeared; it took a generation for it to be absorbed into the technical consciousness of physicists.

When undergraduate physics students first start to work with these equations, there is an understandable "rush" in their minds; all these amazing things revealed by such a small number of symbols. To me, an interesting issue is, Where does the information actually lie? After all, it takes an educational process spanning many years to enable the physicist to work with these equations; is the information in the symbols, or are the symbols a bit like very high level function calls in an API environment, in which they trigger a complex series of mental events that yield the result. The function call without its substrate of lower level operations would be useless. Maxwell's equations without the technical education are just so many hieroglyphics, as they may well seem to CPSC students!

Enter the Electromagnetic Spectrum

Maxwell initially formulated his theory based on an idea of the electromagnetic effect being due to a sort of fluid flow. From the basic equations, he worked out the speed of the interaction; this is a calculation including the two basic physical constants: the electric constant and the magnetic constant. Quantitatively, the velocity comes out to 3 E8 m/s, or the speed of light. This immediately led to the idea that some sort of "electromagnetic" waves could exist, and the possibility that light might be such a wave.

Attempts to detect such waves were carried out by Heinrich Hertz, who demonstrated the first apparatus for creating and detecting low frequency electromagnetic waves, *i.e.* "radio".

In 1902, Guglielmo Marconi demonstrated the transmission of radio waves across the Atlantic. This was the beginning of the radio industry. One of the main applications was the implementation of ship to shore telegraph. Previously, a ship was completely out of touch with the world while on its voyage. The sinking of the Titanic in 1912 was an early example of the use of this technology to summon help.

Gradually, engineers learned how better to control radio waves, allowing for "wave bands" or fixed frequency "bands" allowing simultaneous use of electromagnetic waves by different people in the same area for different transmissions. Today, of course, we have this with our radio stations; we "tune them in" by adjusting our receiver to the desired band.[As a side issue, what about the idea of a fluid for electromagnetism? This idea persisted for some decades, and was finally put to rest by the Michelson-Morley experiment, showing that no physical medium existed (or if it did, it carefully moved so as to nullify any attempt to measure the movement of any apparatus through it!) This led to Einstein's Special Theory of Relativity, and the equivalence of mass and energy, and the atomic bomb.]

Frequency Bands

Early radio technology was of necessity very simple minded, one might say crude. Although high frequencies could carry more information the simple electronic valves (*i.e.* "tubes") could only operate at relatively low frequencies, in the megahertz region. Thus, the first broadcast channels appeared around 1 MHz.

A channel can be thought of as an independent path for information. Modulation is the process of coding information into a channel. A technique called "amplitude modulation" or AM allowed audio information to be coded into a radio frequency channel of narrow bandwidth, in fact only double the frequency range of the audio itself. Since audio goes to 20KHz, this would imply that an AM channel of 40 KHz width would be needed. In fact, it was decided to use a bandwidth of only 10 KHz, since this was adequate for the sound reproduction technology back in the 1920s when radio appeared. [Radio killed off much live theater, vaudeville, and the domestic sale of pianos!] Thus, we had about 100 possible channels in the range 600-1600KHz. A problem with all electromagnetic channels is that in any one reception area, only one channel at any one frequency is usable. Because AM radio stations "carry" quite a long way-the lower frequency ones diffract around the earth's curvature, the higher frequency ones bounce off the ionosphere, especially at night, the 100 channels must be spread around the country fairly thinly. If two channels on the same band are picked up at the same time by a receiver, a squealing noise is heard.

Above the AM band, a number of miscellaneous AM bands developed for various services, collectively called "short wave". The ionospheric bounce was particularly effective in some of these bands, making them useful for military transmissions, and propaganda broadcasts, especially in the "Cold War" from 1950 to 1990. I can remember as a teenager playing with shortwave tuning into Russian rantings (in English) about the "dictatorship of the Proletariat". I can't imagine who would want to listen to this!.

Relatively recently, a band at about 27MHz was allocated for AM Citizen's Band radio. This originally had 24 channels, later expanded to 40. You can see how narrow the CB band looks on the diagram, compared to AM radio. This is due to the logarithmic plot I have used for frequency in the horizontal axis. More modern technology has allowed these higher frequencies, where there is more room for channels, to be used economically, with the aid of "crystal filters" and "frequency synthesis" techniques, not available in the 1920s.

An inventor named E H Armstrong invented many of the basic analog circuits used for radio receivers. A problem with AM radio was its lack of noise rejection. He conceived of the use of another technique, frequency modulation or FM. This would need about 10 times the bandwidth per channel, requring the use of higher frequency bands where there would be enough Hz available. He worked with RCA, who gave him the use of a laboratory and a transmitter setup on the roof of one of the new New York skyscrapers then being built. However, as mentioned in a previous page, Sarnoff, the head of RCA, decided to throw his resources into the development of television. Thus, FM lost the support of RCA, Armstrong committed suicide, and TV ended up with much of the spectrum space. You can see FM squeezed in between the TV channels. RCA threw Armstrong a bone by making the TV audio frequency modulated, but the picture is AM.

Each FM band gets 100KHz, which Armstrong felt was a sell-out; he wanted more. Each TV band takes about 4500Khz, with channels about 6000Khz apart. This separation allowed cheap TV receivers to reliably receive adjacent channels in the same market area.

The long carry of short wave stops working at around 20MHz, so that TV and FM stations are basically "line of sight." This necessitates high transmitting antennas, and allows each major market to have lots of available channels.

In effect, a copy of the electromagnetic spectrum can be carried inside a cable, as an electrical analog, up to frequencies of about 500MHz or so. So, we have cable TV and other services, compatible with the airborne electromagnetic channels.

Finally, at 800 MHz, we find the original cell phone band, used for analog phones.

The next image shows the two digital cell phone bands. I need to find out what bands WI-FI uses, and the upcoming ultra high speed wireless network bands. This latter, I understand, uses "spread spectrum" technology; the information is carried using many different frequencies rather than on single frequencies. This makes interference a serious problem, except that the power is kept very low, reducing range.

Historically, the TV and Cell bands were originally used only for RADAR (Radio Direction and Ranging). This was developed as a crash programme during World War 2. Like many technologies, it was invented by the British but developed elsewhere, in this case the USA. The British invented the "cavity magnetron" vacuum tube, in which high voltage electrons are injected into a very strong magnetic field, causing them to swirl in a curved path, oscillating at the desired gigahertz frequencies. [When this was first demonstrated to the Americans, one of the leading scientists, I I Rabi, said "It's easy-an analog of a whistle" prompting one of the others there to remark sourly "OK, Rabi, how does a whistle work?"] In fact many audio devices and microwave devices are analogous, since the wavelengths are similar. Treble frequencies for audio have mechanical waves similar in length to gigahertz electromagnetic microwaves.

Radar was very crude in its use of the frequencies, basically just an on-off setup. Our modern cell phones are far more sophisticated in their modulation techniques, allowing thousands of channels in any local area.

Apart from crude things like microwave ovens and heat lamps, a huge stretch of the electromagnetic spectrum is currently unused. One reason is that our atmosphere is opaque to these frequencies. This in fact is a cause of the greenhouse effect. The short wave infrared can penetrate the atmosphere, coming from the Sun. It is absorbed by the earth, and re-emitted as long wave infrared. This is because the earth is much cooler than the Sun, and the wavelength produced by an emitter of electromagnetic radiation shortens as its temperature rises.

We generate radio by electrical excitation. Heat will also generate it, although not in a useful fashion, as there is a mess of random frequencies produced. Frequencies like our TV would be characteristic of temperatures very close to absolute zero.

Finally, we reach the visible band, a very narrow strip from about 400 to 800 THz. This band exists because our atmosphere is transparent in this narrow range, and our eyes thus work in this area. Fortunately, the air blocks ultraviolet except the very near UV. This is fortunate because this is the start of the bands that can disrupt chemical bonds.

We are starting to make use of light in CPSC, for fiber optics. Certainly, there are a lot of Hz there. However, our modulation is still rather crude, although it is improving; in fact, unexpected improvements may have contributed to the great internet "bust" in 2001. Companies laid huge amounts of long distance fiber, expecting that each fiber could carry maybe 5 channels. So, methods of sending 200 channels appear, resulting in a vast oversupply of fiber bandwidth! Channels can be sent over fiber using different colors. This is called "wave division multiplexing". I suppose our AM and TV channels could rejoice in the same term!

Visible light in many ways is a "watershed" in terms of wavelength and frequency. The visible band shows both particle and wave characteristics, depending on what experiment is done to observe. Visible light shows wave effects like diffraction and interference, but also shows particle like effects (*e.g.* the photoelectric effect; Einstein got his Nobel from his explanation of this effect, not from relativity).

A whole treatise could be written on the physical causes of visible colors. Basically, they result from energy transitions within atoms, of electrons moving from one state to another. Colors towards violet are caused by higher energy transitions. As we move into the UV, the transitions can involve molecular disruption. The Ozone Layer blocks near UV; if it gets through the degree of molecular disruption on the surfaces of living things becomes an issue, with skin cancer, etc, possibly becoming a serious problem.

Continuing to higher frequencies, X rays are characteristic of very high energy electron transitions. They are produced in CRTs by the electrons slamming with 20000 Volt potential into the inner face of the tube. The tubes are made with lead glass to block these X rays. "Harder" X-rays come from low level nuclear disruption; the nucleus of the atom is about 6 orders of magnitude smaller than the atom itself, and about 6-7 orders of magnitude harder to change in state. Nuclear changes of state (*e.g.* atomic or hydrogen bombs, or the core of the sun) release gamma rays.

There are isolated extremely high energy gamma rays, about 10 orders of magnitude beyond my diagram, occasionally seen. Their source has been a mystery for decades; I believe some tentative explanations now exist.

Shannon's Law and Data Transfer

The Bell Telephone System ran the telephone network in much of the USA until the late 1970s or early 1980s as a government regulated monopoly-almost a public utility. Since they had a protected position, with a guaranteed profit, they could engage in long term research into issues related to communications.

Since by law they could not sell ancillary discoveries for profit, they basically gave away many of their great developments, such as the transistor. [We might note that their Unix developed just as the Bell system was being broken up by the US Government, and the legal restrictions were also being removed. So, Bell started trying to sell Unix licenses, for an amount about equal to the gross national product of Luxembourg, with the predictable result that several non Unix Unices developed, thus "forking" the development of the system, producing several incompatible versions, and thus aiding Microsoft in its march to hegemony.]

1948 was a great year. The transistor was announced to the world, the Manchester University (England) "Baby" electronic computer ran the first stored programme in June, and Claude Shannon announced his communication theory.

Shannon worked much of his professional life at Bell Labs. His main claim to fame is his study into the error-free information capacity of an analog channel. (We mustn't forget as we think about digital technology that it all is really analog underneath). Analog operation involves continuous functions of variables such as voltage, current, time.

Noise is any unwanted energy; it arises from many causes. The most basic is thermal noise caused by the random motion of atoms and molecules in all matter. Noise comes in from the cosmos also; some of the "snow" visible on a TV with an antenna if tuned to a vacant channel may be from this source [Some of this may be echoes of the Big Bang!]. Not surprisingly, human technology produces lots of noise; any electrical switching or sparking produces electromagnetic noise. Computers are of course changing state billions of times per second; each state change is physically done by voltages within the system changing. This will propagate electromagnetic noise. [In fact, there are regulations regarding the amount of RFI (Radio Frequency Interference, or noise) that can be produced by any electronic device. One problem with all this sort of regulation is that compliance is expensive; a small company must send its device to a consultant lab which will run the necessary tests and sign the forms that go to the regulators, a process that may cost $10,000 to $100,000. This process is especially onerous in Europe; where some have cynically suggested that one reason for the stiff regulations is that the big industries, like Siemens and Philips, can maintain staffs for dealing with compliance issues, while small companies face a serious barrier to entry in the marketplace as a result. In North America, the regulations are not so stiff, and in Asia, it is still the "Wild West" in many places.]

Signal to noise ratio is the ratio of the signal value (voltage, current, energy, pressure, or whatever) divided by that of the noise. I will use a value of 1000 in the examples below. This corresponds roughly to AM radio, or the telephone on a good day. High fidelity audio amplifiers can achieve values of around 100,000, limited by thermal noise in the transistors and resistors. [Professionals often measure this in decibels. For voltage or current measurements, the decibel (dB) value can be determined by multiplying the log (10) of the S/N by 20, so a S/N of 1000 would be 20x3 or about 60dB, and 100,000 would be about 100dB.]

Finally, let's consider light. The visible spectrum is about 4 x 10^14 Hz wide. If we use the same 1000 S/N we get a gross data capacity of 4 x 10^15 bits/sec, or about 4 x 10^14 bytes/sec, or about 4 x 10^8 Megabytes per second, or about 4 x 10^5 gigabytes per second, or about 4 x 10^4 DVD movies per second!

Unfortunately, this assumes that we can control light the way we can control radio frequencies, with detailed control over each electromagetic "vibration". This is not possible with light due to things like "Doppler broadening" caused by the thermal motion of the atome producting the light. However, much progress can be made from our current technology.

SATELLITE ORBIT PROPAGATION

Modeling satellite orbit propagation can be accomplished tram a deterministic model or a statistical model of numerous orbital ephemerae. Because the deterministic approach is so well known in classical physics, we do not seek to repeat the equations and algorithms. While fundamental in nature, these equations would still take some ten pages or so for presentation and more for explanation. The deterministic approach is well grounded but complex (Hoots and Roehrich 1980). One of the earliest works on deterministic modeling of satellite orbits around the Earth was by Brouwer (1959). A brief history of the fundamentals of Brouwer's research and the subsequent modifications by others is presented by Hoots (1981). Tapley et al. (2004) provide a thorough discussion of statistical approaches for modeling satellite orbit propagation. Notable works by others (e.g., Emery et al. 1989) for specific sensors on remote sensing satellites have also been conducted. Those interested in developing their own satellite propagation or sensor models are encouraged to consult these fundamental works.

The orbital propagation solution and sensor viewing geometry developed for the Remote Sensing Hazard Guidance System (Hodgson and Kar 2008) was used in this research. We have built our model based on these works but modified it for efficiency and accuracy. The age of the satellite ephemerae (i.e., the location and trajectory of a satellite observed at one moment) is one of the major factors influencing satellite orbit propagation accuracy. Predictions of near-future satellite locations (e.g., hours later) are trivial and very accurate, while predictions for months ahead require greater complexity and numerous additional factors. Our model was based on other demanding applications but was used in a very ideal context here. For this research, we used orbital ephemerae that were collected nightly and thus are less than 24 hours old. Our orbital predictions for day 'n' used the ephemerae from day 'n'-1 (i.e., the preceding day). By using daily ephemerae for predictions within the next 24 hours at the most, we also obviate the major sources of error in satellite orbit prediction, such as satellite maneuvers.

The first step in the orbital propagation process is to obtain the orbits of all the satellites of interest. The orbital ephemerides are obtained on a nightly basis for all remote sensing satellites. The satellite orbits are converted into a Cartesian Earth centered state vector at the particular epoch of interest. The Cartesian state vectors are used to compute classical orbit elements for each of the satellites. These classical orbit elements may be used to propagate the satellite trajectory forward or backward in time from the epoch specified. In addition, the time and state vector for a particular satellite may be determined using angles from the satellite location at the epoch of interest. For example, the state of a satellite when it reaches its northern most position may be determined from the angle measured from the initial position to the point in the orbit where the true anomaly plus argument of periapsis is 90 degrees. Using these classical orbit elements, the parts of the orbit where the spacecraft is on the dark side of the Earth or too far north to be imaged may be quickly eliminated.

The search algorithm then finds the point on the orbit where the satellite is closest to the landfall point defined on the surface. At this point, the range rate is zero and transitioning from negative to positive. These geometrical parameters may be easily computed from the landfall point location and state vectors of the satellite obtained from orbit element propagation. Various geometrical parameters, such as sun elevation angle or range, may be examined to determine if they satisfy the constraints specified. If they are out of tolerance, the satellite will not have an opportunity to collect imagery at this location.

Once the minimum-range point is found, as determined from conic orbit elements, a new initial epoch is computed. The ephemerae database is interrogated for a new satellite state vector for the date/time calculated using the conic orbit element. The entire search algorithm is then repeated using precision satellite state vectors at the exact time of overflight. This insures that the most accurate results are obtained, and conic orbit propagation is only used for relatively short distances.

Satellite Viewing Opportunities

To empirically determine the likelihood of future collection opportunities, the spatial and temporal distribution of historic landfall points were assumed to occur in the year 2005. The year 2005 is arbitrary but was convenient for this research as we had a complete daily collection of the satellite orbital ephemerae needed to predict the orbital tracks of each satellite and, subsequently, the satellite viewing opportunities. As noted earlier, using daily orbital ephemerae allows for very high spatial/temporal accuracy. The use of any other year may change the collection opportunities for a single hurricane event but not for the set of landfall events, unless the orbits of the satellites were temporally correlated.

For each day and geographic landfall location of a hurricane, the viewing opportunity for each of three satellites (i.e., Ikonos-2, Quickbird-2, and Orbview-3) was modelled. Determining if the geographic location can be imaged is solved by first propagating the satellite vehicle orbit during the days following the hurricane landfall, and modeling the viewing geometry for the sensors onboard the satellite vehicle. Most satellites with high spatial resolution sensors have a similar orbital period (90-100 min) resulting in similar repeat intervals on the ground (satellite image providers routinely advertise a 2-3 day repeat visit).

Imaging opportunities vary among satellite sensors with variations in altitude and off-nadir viewing. The OSA sensor on Ikonos-2 and Orbview-3 can point 45-degrees off-nadir as compared to the 28-degrees of the BGIS200 on Quickbird-2. For most mapping applications (except for stereoviewing applications), off-nadir viewing angles of less than 26 degrees are desirable. At greater angles, the ground surface obscured by the sides of land cover features (buildings, vegetation, etc.) is problematic. The potential swath coverage depicts the possible image collection opportunities; however, it does not represent the area-field-of-view, also called swath width, which would be imaged by most high spatial resolution sensors. The swath width is always smaller (e.g., 16.5-km in width for Quickbird-2 at nadir) than the potential swath coverage.

The likelihood of satellite imagery collection was examined under several scenarios. First, we assumed a single image provider, and a maximum viewing angle of 26 degrees off-nadir was permitted. Historically, remote sensors in a hurricane disaster context have sought imagery from near-nadir views rather than oblique imagery. The 26-degree criterion has been used informally in emergency operations centres. Second, any of the three satellites were allowed but still imposing a 26-degree off-nadir limit. Finally, collection opportunities from 1) single satellites and 9) all three satellites were examined with the maximum viewing angle allowed by the sensors.

Restricted Viewing Angles

If an emergency response agency relied on a single satellite provider to collect imagery over the hurricane landfall area, the likelihood of obtaining imagery within the first 24 hours ranges from 16 to 18 percent. The likelihood increases to 37 to 42 percent during the first 48 hours and from 53 to 67 percent in the first three days. Even with the restriction of 26 degrees maximum viewing angle, the Ikonos-2 satellite-sensor combination has a higher likelihood of collection as its altitude is higher (i.e., 679 km versus-450 km for the other two).

A higher altitude allows the sensor to point at farther planimetric distances with the same viewing angle. Relying on one satellite-sensor combination with this 26 degree off-nadir viewing limit results in an image collection during the first three days of, at best, only 2 out of 3 (-67 percent) hurricane disasters. However, if satellite imagery from any of the three satellites were permitted, then the likelihood increases to 38 percent, 74 percent, and 91 percent within the first 24, 48, and 72 hours after hurricane landfall.

Maximum Viewing Angles

Not surprisingly, when more extreme viewing angles are allowed (e.g., when the agency must have some imagery regardless of whether portions of the landscape are obscured), the likelihood of collecting imagery increases. Ikonos-2 and Orbview-3 can point off-nadir at angles of 45 degrees. The likelihood of collection in the first 24 hours increases to 39 percent using either of theses satellites. The likelihood increases to 97 percent for either Ikonos-2 or Orbview-3 within the first 72 hours after landfall. By relying on all three satellite image providers, the likelihood of collecting in the first 24 hours is 61 percent.

This research estimated the likelihood of using one or more (up to three) satellite imagery sensors for collecting post-hurricane event imagery in the first 24, 48, and 72 hours of landfall. Relying on a single satellite provider would result in only a 39 percent chance of collecting in the first 24 hours, if using maximum off-nadir viewing angles. To improve the collection likelihoods for post-event imagery, multiple satellite providers must be relied upon. In the concept of operations (CONOPS) model, this would inevitably require pre-existing contracts (pre-existing to enable a rapid agreement to collect and pay for the collection) with multiple satellite image providers. Since the completion of this research, Orbview-3 satellite imaging sensor has malfunctioned and is considered inoperable. However, GeoEye-1, with a.41 cm spatial resolution and a 681-km altitude and off-nadir viewing angle of up to 61 degrees is operational. Using the GeoEye-1 satellite in lieu of Orbview-3 would result in likelihood estimates slightly better than Ikonos-2. Other high spatial resolution satellite-sensors, such as Worldview-1, have recently been launched, providing viewing opportunities from at least four different U.S. commercially controlled satellites.

A logical planning scenario might be to use multiple satellite-sensor contracts and stratify the potential damage area into image collection priorities. An image collection with a near-nadir (low off-nadir angles) vantage might be used for the high-priority areas, while greater off-nadir angle opportunities are used for low priority areas.2 Focusing on large-impact areas would further increase the likelihood of collecting imagery across a disaster area.

A final probing scenario for response agencies might be to use both aerial and satellite imaging platforms in a well defined and coordinated approach. By quickly modeling (e.g., GIS-based without imagery the likely damage area, the collection area could be spatially defined, stratified, and prioritized. Available airborne and satellite-sensors could be utilized in a stratified (both temporally and spatially) approach to collect imagery as part of the post-event response.

There is a revolution underway in Caddo archaeology, and its name is remote sensing. The use of remote sensing technologies on Caddo archaeological sites-in conjunction with archaeological excavations-is producing, and will continue to produce, unprecedented characterizations of the internal spatial structure and organization of Caddo villages and mound centres. We illustrate the importance of remote sensing for the present and future of Caddo archaeology by focusing on the remote sensing and archaeological findings from one Caddo village on the Red River in Bowie County, Texas. The Hill Farm site is part of the larger Hatchel site village and was occupied by the Nasoni Caddo between ca. A.D. 1600 and 1700.

The Nasoni Caddo village was visited by Europeans in 1687, 1690, and 1691, most famously by the Spanish expedition led by Don Domingo Teran de los Rios. On this expedition (Hatcher 1932, 1999), a detailed map of this village was drawn that showed a temple mound and a number of village compounds (with house structures, granaries, and outdoor ramadas or arbors). A convergence of large-scale remote sensing (Grealy and Conyers 2006), the smaller-scale remote sensing, geoarchaeological findings (Guccione and Hays 2006), and the archaeological discoveries at the Hill Farm site and other places at the Hatchel site (Perttula 2005) have led us to the specific identification of the Hill Farm site with two of the village compounds depicted on the 1691 Teran map. This in turn opens the door to the possibility that continued archaeological and remote investigations can, in effect, locate and ground truth the 1691 Teran map and different parts of the Nasoni Caddo village.

Remote Sensing in Caddo Archaeology

Archaeogeophysical prospecting using magnetometry, electromagnetic conductivity, electrical resistivity, ground-penetrating radar, and magnetic susceptibility have all been shown to be useful geophysical techniques for locating buried architectural remains on prehistoric Caddo mound sites, such as the George C. Davis site and Horace Cabe Mounds (Walker and Perttula 2007) in northeastern Texas, and at various sites on the Grandview project in southwestern Arkansas. These techniques measure certain physical properties of the soil, such as magnetic conditions, electromagnetic conductivity, or electrical resistivity. Given the proper soil conditions, archaeological features such as fire hearths, post holes, and storage pits differ from the surrounding soil matrix, and can be recognized by distinctive spatial signatures in geophysical data sets from archaeological sites.

At the George C. Davis site, geophysical surveys, using a portable cesium magnetometer, on large blocks of land have proved successful in locating both large and small architectural features, particularly numerous Caddo structures. This work is contributing new information on the overall layout of structures around the mounds and across this large prehistoric site.

In some settings, the geophysical survey has successfully located archaeological features in areas that previous subsurface excavations failed to locate (Walker et al. 2003). Shaded relief maps of the magnetic data have also identified a number of unique architectural forms at the site, such as the so-called Button Houses (Creel et al. 2005), that can be compared to the architectural plans of the more than 50 previously excavated structures at the George C. Davis site.

These geophysical examples of Caddo structures are 10-15 m diameter circular structures with four internal magnetic returns that are spatially patterned on the interior of the structure to encircle a central hearth; these no doubt represent large post hole features (support posts) or interior pits. The magnetometer data from the Mound B area at George C. Davis are providing new information about the construction phases of the mound. These data have detected a trench of some kind or a small mound of earth that apparently served to anchor the mound to a specific place on the ground. The remote sensing work has also identified a second ramp on the southern flanks of the mound (Schultz et al. 2004).

On the Grandview project, the Arkansas Archeological Survey used a variety of geophysical surveying equipment to measure magnetic susceptibility, gradiometry, electrical resistance, and electromagnetic conductivity of key areas at Caddo mound and nonmound settlements, and achieved impressive results. The geophysical surveying has detected the locations of burned Caddo houses, fireplaces or hearths within the houses, burial pits, nonburial pits of various sizes and shapes, stockade lines around the Tom Jones mound (3HE40) (Lockhart 2007), and possible paths across the sites.

Ground-truthing excavations of several of the subsurface anomalies identified in the Grandview geophysical survey have been successful in identifying archaeological features, including burned houses with "heaps of fired daub up to a meter deep in places, pieces of carbonized logs, and a floor laden with ash and domestic debris" (Schambach 2001:10). Other excavations exposed clay-lined hearths, and lines of post holes, while excavations placed over "reasonably clear images [from the remote sensing] of house remains netted... the archaeological remains of four houses" (Schambach 2001:10).

The Archaeology of the Hill Farm Site within the Larger Hatchel Site Village

Before discussing the remote sensing findings and our interpretation of them, we review what is known about the archaeology of the Hill Farm site. The site is only one part of the much larger Hatchel site village occupied by the Nasoni Caddo from the eleventh to the eighteenth century (Perttula 2005).

The village was apparently comprised of a number of farmstead compounds spread out over many acres, with domestic structures, granaries, arbors, and ramadas. The farmstead compounds may have been separated by compound fences or dividers of trees and bushes. The Hill Farm site has two areas, Areas A and B, with substantial archaeological deposits in both.

The first archaeological investigations at the Hill Farm part of the Hatchel village was completed by the Works Progress Administration (WPA) and the University of Texas at Austin in 1939 (TARL n.d.a). The work was on a small Caddo mound, 21 m in diameter and 0.9 m high, which is 915 m southeast of the main platform mound at the Hatchel site and 610 m southeast of WPA excavations in Village Plot 1 (Perttula 2005).

The mound (Mound 2) is in Area A at the Hill Farm site; its exact location is currently unknown. Three other small mounds (0.6-0.9 m in height) were nearby. Just east of Mound 2 was a borrow pit of unspecified size (TARL n.d.a).

The WPA excavations in Mound 2 were in two blocks at the northern and southern parts of the mound.

The northern strip covered 150 ft^2 and the southern excavation was 900 ft^2 in size. These excavations disclosed that the mound was constructed in one episode of a brown clay fill that was 0.9 m thick. It was probably constructed to bury a structure or important public building, although the details of this building are not known. The mound fill was placed atop a 25 cm thick habitation and midden deposit of Late Caddo Texarkana phase age (ca. A.D. 1400-1700). No structural features or post holes were recorded in these submound deposits.

A single extended burial (Burial 1) was excavated in the southern Mound 2 strip. This adult burial originated at the top of the habitation/midden deposits-prior to mound construction-and the grave pit reached to 1.37 m below surface (bs).

No funerary offerings were placed with this Caddo individual. There were seven sherds, including a vessel lug, in the burial pit fill. Only 324 artifacts were recovered in the Mound 2 WPA excavations (none of the archaeological deposits were screened during the WPA work). This included 212 ceramic vessel sherds, one celt, 25 pieces of lithic debris, 71 animal bones, and 15 pieces of freshwater mussel shell.

In the 1950s, according to records on file at the Texas Archeological Research Laboratory, University of Texas, two avocational archaeologists (R. King Harris, from Dallas, Texas, and Pete Miroir, from Texarkana, Texas) conducted excavations in what appears to have been Area B at the Hill Farm site. They apparently encountered several Caddo burials and very large post holes from a large structure. No more details on this work are available, unfortunately.

More recently, Gilmore and McCormick reported that an unknown number of Caddo burials had been looted at the Hill Farm site in the 1970s. Their location within the site is also unknown.

In 2005, renewed archaeological investigations at the Hill Farm site were completed as part of the Bowie County Levee Realignment project sponsored by the Tulsa District of the U.S. Army Corps of Engineers.

During this work (Sundermeyer et al. 2006), 78 shovel tests were excavated in Area A (27 of which contained prehistoric archaeological materials), with another 127 shovel tests (51 contained prehistoric archaeological remains) in Area B.

Additionally, a magnetometer survey was completed in both Area A (6,000 m^2, ca. 1.5 acres) and Area B (3,600 m^2, 0.9 acres) by Walker and Schultz (2006). It is the findings from both the shovel testing and the remote sensing that we discuss in the remainder of this chapter.

The Hill Farm site contains exceptionally well-preserved and buried Late Caddo Texarkana phase archaeological deposits. The archaeological materials are incorporated in Red River alluvial sediments of the abandoned and recently defined Cemetery Meander (Guccione and Hays 2006). These deposits apparently began to accumulate sometime after ca. A.D. 1240.

The site lies on sandy loam point bar alluvial deposits adjacent to an ancient abandoned channel of the Red River.

The archaeological deposits are marked by numerous structures of varying sizes, pit features, midden deposits, at least one constructed earthen mound (Mound 2 at Hatchel) and as many as three other small mounds, a borrow pit, and one extended burial.

The recent shovel testing reported by Sundermeyer et al. (2006) identified two distinct areas (A and B) at the Hill Farm site. Area B, the northern area, covers approximately 13,000 m^2 (3.2 acres), while the southern part of the site (Area A) is about 11,000 m^2 (2.7 acres) in size.

The two different site areas are about 160 m apart. Archaeological materials occur to depths of 80100 cm bs. The vast majority of these remains are buried, coming primarily from 20-80 cm bs. There are two concentrations of ceramic vessel sherds in Area A, with one large and heavy-density cluster covering ca. 2,400 m^2 in the centre of this village locality.

A much smaller and lighter scatter of sherds is ca. 15-30 m south of the principal sherd concentration. In Area B, there are 4-5 distinct concentrations of ceramic vessel sherds. These concentrations are 15-40 m apart from each other within this village area, and most of them are situated in the southern part of Area B.

4

Remote Sensing Processing and Geological Structures

The interest in 3D reconstruction comes from the intention of modelling the geological reality in a very faithful way, offering many advantages over 2D representations. The construction of 2D geological sections involves several steps, including the selection of data to project onto the section plane, the orientation of the section plane, the projection distance, and the orientation of a projection vector of regional significance.

All of them may generate unexpected errors, which are difficult to quantify or anticipate because of their subjective nature (Fernández et al. 2003). When reconstructing geological structures directly in three dimensions we attempt to avoid the various simplifications and their related errors during cross-section construction. To adequately represent the geometrical variability of 3D geological-structural geometries, a uniform distribution of data along the structure is preferable. However, the availability and distribution of the dataset can be limited or scarce in regions with difficult access because of geographical or geopolitical reasons. That is the case of the study area, the Anaran anticline, located along the Mountain Front Flexure in the most external part of the Lurestan province.

This anticline is characterized by strong geometrical changes along strike, thus making it difficult to solve by 2D sections. To overcome the difficulties of 3D structural reconstruction in areas with difficult access or where data availability is scarce, we present a methodological workflow for 3D reconstruction of geological surfaces, at a regional scale, using remote sensing data and geological maps. The reconstruction technique is based on the 'dip-domain' concept, introduced in the first kink-style structural models.

Following this approach, any geological structure can be discretized in constant dip and azimuth volumes, bounded by surfaces with geological significance (faults, axial planes, unconformities). The application of this concept to 3D geological reconstruction was initially developed by Fernández et al. (2004) for a local-scale case study and with a dense dataset mainly derived from field studies, 3D mapping at outcrop scale and subsurface data that extensively represented the structure at the reconstruction scale.

When dealing with a regional case study without field data support or with poor field data, we have to have recourse to additional methods to increase the density of the dataset whenever it is considered to be too scattered or discrepant. An example is the use of plunge lines to project data, as well as the local use of automatic interpolators.We combined in the same workflow the gathering, storing and processing of spatially referenced information with the geological constraints, the structural analysis and the 3D elevation model necessary for the 3D reconstruction. The approach presented here is based on the equivalent dip domain method recently proposed by Carrera et al. (2009).

Methods

The workflow has been subdivided into three main steps, each of them following specific methods. The first step (remote sensing processing) consists of the integration, unification and improvement of the remote sensing data coming from various sources. In the second step (3D mapping and structural analysis), the resulting geological map is analysed to produce a structural and stratigraphic model. In the third step (surface reconstruction) the selected horizon is built by applying the dip-domain concept.

Dataset

The initial dataset used for the 3D reconstruction is: (1) 1:100 000 Geological maps prepared and published by the Oil Service Company of Iran; (2) Spot 5 images: 2.5 m per pixel resolution in the panchromatic channel with a swath size of about 60 km 3 60 km; (3) Aster images from L1A level (Reconstructed Unprocessed Instrument Data), which are mainly raw data; the swath size is about 60 km × 60 km and the spatial resolution ranges from 15 m to 90 m per pixel; (4) a Spot digital elevation model (DEM), gridded with step sizes of 1 s of arc 30 m on the equator (east-west dimensions vary according to the latitude) resampled to 20 m. The absolute horizontal accuracy is estimated with 90% confidence (according to the provider) with an error level of less than 15 m, whereas the absolute vertical accuracy, considered as the DEM elevation error relative to ground truth, depends mainly on the slope of the terrain. If we consider only slopes below 20% the positioning error is no greater than 10 m, but with greater values of slope the vertical error can increase considerably. As a consequence, cliffs' shapes become smoothed and are no longer representative of the real morphology.

PASSIVE REMOTE SENSING

Passive systems collect data from energy that is reflected or radiated off the Earth's surface and atmosphere. A typical image derived from an infrared passive sensor consists of small equal areas referred to as pixels (7) arranged in regular rows and columns. Each pixel has a numerical value called a digital number (DN) that records the intensity of electromagnetic energy measured for the area of ground represented by the pixel. The DN range from 0 to some higher number on a gray-scale. Each pixel is also given x and y coordinates to place it. The image can therefore be described in strictly numeric terms on a three-coordinate system with x and y locating the Pixel and z giving the DN displayed as a gray scale intensity value.

Passive sensors are described in terms of their spatial, spectral, and temporal resolutions. The spatial resolution of a sensor is the smallest area that is recorded as a separate unit (pixel). For instance, one-meter spatial resolution means that one pixel of a digital image represents an area on the Earth's surface measuring one meter in length by one meter in width. Spectral resolution refers to the number and dimension of bands (or wavelengths) of the electromagnetic spectrum that a sensor records. The higher the number of bands, the greater the sensor's ability to distinguish between objects. Temporal resolution, also known as repeat time, is the frequency with which a sensor passes over the same area.

Active Remote Sensing

Active remote sensing devices, on the other hand, emit high-energy electromagnetic radiation and record the relative amount and pattern of the energy that is reflected back. Many of these devices operate at wavelengths that not only penetrate cloud cover, but also vegetative cover and soil surfaces. The tradeoff for greater imaging capabilities, however, is increased complexity in data interpretation, as compared to passive sensor data interpretation.

Data Processing

After the satellite records the data, it is transmitted to a ground station for calibration and storage. The data may undergo various levels of processing before it is made available to the user. These levels range from simply correcting for transmission errors to performing advanced correction and analysis with model algorithms, depending on the needs of the scientists or user.

Once the data has undergone initial processing techniques, users may apply it for various purposes, from the simple production of an enhanced image to the more complex creation of image maps, thematic maps, and spatial databases. The data may also be used to develop statistical observations and graphs of the observed phenomena. To create maps and spatial databases, the initial data must be combined with other spatial data. An effective method to analyse the remote sensing data with reference to other spatial data is in a geographic information system (GIS).

Remote Sensing Data Integration with Geographic Information Systems (GIS)

Geographic information systems (GIS) are defined as computer systems capable of assembling, storing, manipulating, and displaying geographically referenced information (i.e. data points identified with respect to their location). GIS store information about the world as a collection of thematic layers that can be linked together by geography.

Remote sensing data applications and GIS have an established history of interdependency. GIS provides a format to distribute remote sensing data and to derive useable information from the data. Remotely-sensed data is also a critical means to create base GIS maps and update many data layers in the GIS. The integration of remotely-sensed data and GIS is particularly attractive because 1) the conversion of remotely-sensed raster-format data to GIS vector-format data is inexpensive and 2) remote sensing data offers a cost-effective way to visualize large geographic areas in a digital format.

There are two defining features of all GIS: the ability to overlay spatial data and the ability to change as new data becomes available. The first key feature of GIS programs is the capability to overlay multiple sets of databases into a map format that graphically explains the relationships between the data. Spatial data (points, boundaries, and lines) comprise the base of the map and can be supplemented with tabular data and image data (such as that from satellites). This powerful and versatile concept has proven invaluable for solving many real-world problems, from recording details of land use planning applications to modeling global atmospheric circulation cycles. The second key feature of GIS is their status as "dynamic maps" that can be updated and altered as needed. These maps may also be manipulated to perform scientific analyses and to create models of different environments.

In a simplistic example of GIS application, a map of city streets could be combined with latitude/longitude-referenced traffic flow data to create a map that reveals areas of frequent accident occurrence, potential detour routes, and even alternatives to improve traffic routing and alleviate rush hour stress. The same base map also may be reused to show, for example, changes in traffic patterns across time.

DEVELOPING OPEN DATA, OUT OF PLACE-BASED MAPPING

The voice of the participant, rather than the voice of the researcher, will be heard best when participants not only provide the data to be analysed, but when they also contribute to the questions that frame the research and contribute to the way data are analysed.

In terms of mapmaking praxis, construction of the map has been an entirely de-centred process and authoritative models of data collection and transcription have been overturned. The application of ethnographic methodology ensures multiple voices construct the map. Within this work the relationship of emotion, memory, and sensory engagement with the landscape was mapped. First, data was sourced while walking, talking and recording with participants on the landscape. After an initial recruitment period and focus session, each co-developer chose a location for a 'one-to-one' walk that in some way was connected to the lake. Co-developers chose the date and time – some brought their binoculars or dog along, others even brought 'somebody else along'. The co-developers were helped in tracking the walk; sites of interest, objects, plants and animals, favourite places, memory spots and stories connected to the place. Places were noted using cameras, notebooks, a GPS unit and a dictaphone. A framework of open and closed questions was asked. Answers to open questions, such as 'What do you feel about the lake?' were geo-tagged. The 'type of walk' (their special walk) became an integral part of data collection and data analysis. These processes helped capture the walk and created a supplementary resource to each recorded conversation that took place, which was then transcribed.

In the spirit of ethnographic methodology, transcription and coding of data was a mutually inclusive activity (all information was verified with co-developers). Some of the codes that developed from the walks were words such as rocks, water, agriculture, birds, meditation, trees, fields, memories, fish and events. In a paper-based workshop, co-developers then jointly discussed the codes and each shared their record of the walk from memory. Memories were added to the discussed codes. Importantly here, qualitative data became coded by the co-developers and not by some distant and 'removed' ethnographer. To this end, paper-based tags were ready-made for the map interface. A sort of starter kit had been created, effectively introducing co-developers to tagging or folksonomy. Qualitative coding methodologies in turn introduced the community to the art of good folksonomy. This is an important issue, because it deepened the practice of folksonomy and helped to reflect on it in a practical and academic manner.

A computer workshop then tested the beta version of the sustainable software. Each co-developer put marks on the map, using latitude and longitude figures supplied from the archive of walks and paper-based workshop.

They tagged their marks efficiently and with ease, having been introduced to the concept of folksonomy in the paper workshop. When things started taking shape onscreen the mood in the workshop room was electrifying. Everyone watched their places appear on the map – and all the efforts and concepts that must at times have seemed utterly puzzling started to make sense and finally paid off.

Technical problems were fixed as and when they arose. Co-developers' views, feelings and ideas for the future of the software, as well as ideas for new data, were taken into account. Outdoor events have since become interchangeable with ongoing paper and computer workshops. All types of place-based mapping happenings have been called for by the co-developers, and are enabling the gathering of data that was not pre-determined. For instance, moth migration nights, stargazing gatherings, butterfly balls, drawing picnics and plastic bag counts have taken place. The geo-coded data is challenging how the base map could look and function, and is drawing in experts in the field of flora and fauna and qualitative research.

At present, when data is added, co-developers either leave trails of red dots where they have been recording a walk or they add to pools of information where groups have gathered. For instance, a moth and bat night focused on three spots at the mouth of the lake, and became like three micro-maps of fascinating creatures and facts. These red marks could become a sea of pictures, telling a tale of moths in this area. At the moment pictures are uploaded to a separate space, to a group Flickr (2004) account. As funding is applied to develop the project further, the co-developers will become involved in the application, asserting what they think should be developed next. Their priorities so far are: to make the base map more sophisticated; to make their map entries about each place more editable; to style the forum and to be able to guest blog (at the moment the project initiators and a few others are blogging).

MAPS, REMOTE SENSING AND GIS

A map can be simply defined as a graphic representation of the real world. This representation is always an abstraction of reality. Because of the infinite nature of our Universe it is impossible to capture all of the complexity found in the real world. For example, topographic maps abstract the three-dimensional real world at a reduced scale on a two-dimensional plane of paper.

Maps are used to display both cultural and physical features of the environment. Standard topographic maps show a variety of information including roads, land-use classification, elevation, rivers and other water bodies, political boundaries, and the identification of houses and other types of buildings. Some maps are created with very specific goals in mind. A weather map showing the location of low and high pressure centers and fronts over most of North America. The intended purpose of this map is considerably more specialized than a topographic map.

The art of map construction is called cartography. People who work in this field of knowledge are called cartographers. The construction and use of maps has a long history. Some academics believe that the earliest maps date back to the fifth or sixth century BC. Even in these early maps, the main goal of this tool was to communicate information. Early maps were quite subjective in their presentation of spatial information. Maps became more objective with the dawn of Western science. The application of scientific method into cartography made maps more ordered and accurate. Today, the art of map making is quite a sophisticated science employing methods from cartography, engineering, computer science, mathematics, and psychology.

Cartographers classify maps into two broad categories: reference maps and thematic maps. Reference maps normally show natural and human-made objects from the geographical environment with an emphasis on location. Examples of general reference maps include maps found in atlases and topographic maps. Thematic maps are used to display the geographical distribution of one phenomenon or the spatial associations that occur between a number of phenomena.

MAP PROJECTION

The shape of the Earth's surface can be described as an ellipsoid. An ellipsoid is a three-dimensional shape that departs slightly from a purely spherical form. The Earth takes this form because rotation causes the region near the equator to bulge outward to space. The angular motion caused by the Earth spinning on its axis also forces the polar regions on the globe to be somewhat flattened.

Representing the true shape of the Earth's surface on a map creates some problems, especially when this depiction is illustrated on a two-dimensional surface. To overcome these problems, cartographers have developed a number of standardized transformation processes for the creation of two-dimensional maps. All of these transformation processes create some type of distortion artifact. The nature of this distortion is related to how the transformation process modifies specific geographic properties of the map. Some of the geographic properties affected by projection distortion include: distance; area; straight line direction between points on the Earth; and the bearing of cardinal points from locations on our planet.

The first two-dimensional projection shows the Earth's surface as viewed from space. This orthographic projection distorts distance, shape, and the size of areas. Another serious limitation of this projection is that only a portion of the Earth's surface can be viewed at any one time.

The second illustration displays a Mercator projection of the Earth. On a Mercator projection, the north-south scale increases from the equator at the same rate as the corresponding east-west scale. As a result of this feature, angles drawn on this type of map are correct. Distortion on a Mercator map increases at an increasing rate as one moves toward higher latitudes. Mercator maps are used in navigation because a line drawn between two points of the Earth has true direction. However, this line may not represent the shortest distance between these points.

The Gall-Peters projection was developed to correct some of the distortion found in the Mercator system. The Mercator projection causes area to be gradually distorted from the equator to the poles. This distortion makes middle and high latitude countries to be bigger than they are in reality. The Gall-Peters projection corrects this distortion making the area occupied by the world's nations more comparable.

The Miller Cylindrical projection is another common two-dimensional map used to represent the entire Earth in a rectangular area. In this project, the Earth is mathematically projected onto a cylinder tangent at the equator. This projection in then unrolled to produce a flat two-dimensional representation of the Earth's surface. This projection reduces some of the scale exaggeration present in the Mercator map. However, the Miller Cylindrical projection describes shapes and areas with considerable distortion and directions are true only along the equator.

The Mollweide projection improves on the Robinson projection and has less area distortion. The final projection presented presents areas on a map that are proportional to the same areas on the actual surface of the Earth. However, this Sinusoidal Equal-Area projection suffers from distance, shape, and direction distortions.

MAP SCALE

Maps are rarely drawn at the same scale as the real world. Most maps are made at a scale that is much smaller than the area of the actual surface being depicted. The amount of reduction that has taken place is normally identified somewhere on the map. This measurement is commonly referred to as the map scale. Conceptually, we can think of map scale as the ratio between the distance between any two points on the map compared to the actual ground distance represented. This concept can also be expressed mathematically as:

$$\text{Map Scale} = \frac{\text{Map Distance}}{\text{Earth Distance}}$$

On most maps, the map scale is represented by a simple fraction or ratio. This type of description of a map's scale is called a representative fraction. For example, a map where one unit (centimeter, meter, inch, kilometer, etc.) on the illustration represents 1,000,000 of these same units on the actual surface of the Earth would have a representative fraction of 1/1,000,000 (fraction) or 1:1,000,000 (ratio). Of these mathematical representations of scale, the ratio form is most commonly found on maps.

Scale can also be described on a map by a verbal statement. For example, 1:1,000,000 could be verbally described as "1 centimeter on the map equals 10 kilometres on the Earth's surface" or "1 inch represents approximately 16 miles".

Most maps also use graphic scale to describe the distance relationships between the map and the real world. In a graphic scale, an illustration is used to depict distances on the map in common units of measurement. Graphic scales are quite useful because they can be used to measure distances on a map quickly.

Maps are often described, in a relative sense, as being either small scale or large scale. We have maps representing an area of the world at scales of 1:100,000, 1:50,000, and 1:25,000. Of this group, the map drawn at 1:100,000 has the smallest scale relative to the other two maps. The map with the largest scale is map C which is drawn at a scale of 1:25,000.

RESEARCH DESIGN : DATA AND STUDY AREA

We applied our technique to an IKONOS sensor image data of 4m spatial resolution across all four of its channels: blue (0.45-0.52 [micro]m), green (0.52-0.60 [micro]m), red (0.63-0.69 [micro]m), and near infrared (0.76-0.90 [micro]m).

The image represents the settlement of Norman, Oklahoma and was captured on March 20, 2000. Only a subset of this IKONOS sensor image (614 x 447 pixels) covering a central portion of the metropolitan area was used to identify three land cover/land-use classes: grassland, commercial, and residential—all three capable of being delineated using manual interpretation.

Local Window

At the onset, it is important to note that the characteristic scale (Lark 1996) is an important parameter to be considered for effective identification of land covers with different texture appearances. A characteristic scale is the minimum distance between two pixels that completely covers a texture. In our study, anything between 21 pixels (84 m) and 27 pixels (108 m) was considered large enough to cover textures that represent all of the classes, especially complex residential and commercial land uses derived from the IKONOS multispectral image. Hence, 9 x 9, 15 x 15, 21 x 21, 27 x 27, 33 x 33, and 39 x 39 local window sizes were used to determine the optimum scale with which to identify land-cover classes.

From a previous study (Myint and Lam 2005b) that was based on a lacunarity approach designed by Voss (1986), we demonstrated that smaller gliding boxes have more discriminatory power than larger gliding boxes. And as a result, a small 3 x 3 x 3 gliding box is used in this study and is the very basis for the chosen window sizes dimensionalized by a factor of 3. Hypothetically, a window size should be small enough to cover only single land-cover features but large enough to guarantee sufficient spatial/textural information for the characterization of land-cover types.

In that case, a mirror extension of (w-1)/2 pixels around the image is necessary before beginning computation. What happens is that the algorithm is designed to extend automatically the image with (w-1)/2 pixels all around if the selected window size is w. As such, the size of an extended image is the original image size + (window size-1). Mirror extension is designed to copy the second-last row or column and add the next-to-last row and column, respectively. Then copy the third-last row and column and add the next-to-the-second-last row and column, respectively, and so on depending on the number of rows and columns required for the extension.

This mirror extension procedure is demonstrated by a hypothetical image and is considered more effective than other widely used methods, including those adding zero, one, last row/column data, or the mean values in the extended areas. Our favoured approach is also more accurate than other interpolation methods (e.g., kriging, inverse distance) because it does not alter any values in the original image. By using the mirror extension, we obtain the same mean and standard deviation statistics within the extended area.

Training Samples

With knowledge of the local area and ground checks, training samples were carefully selected to represent the three land-cover/land-use classes: grassland, commercial, and residential—descriptive statistics of their brightness values are shown. It is generally suggested that the standard deviation of training data should be around, or even less than, 10% of the mean coefficient of variation to guarantee homogeneous distributions vital to traditional spectral-based classifiers. Compared with the standard deviation, which is expressed in absolute terms, the coefficient of variation is considered a more appropriate measure for the comparison of data distributions across variables.

The standard deviation of the commercial and residential land uses is 37.1% and 34.3% of their respective mean values. At this stage, it is important to remember that groups of adjacent pixels generally exhibit positive autocorrelation, or at least display a high probability of having similar digital numbers. In our calculations, the grassland (Moran's l = 0.49; Geary's C = 0.43) sample exhibited a positive autocorrelation, whereas the commercial (Moran's l = 0.16; Geary's C=0.83) and residential (Moran's l = 0.09; Geary's C = 0.91) samples displayed random spatial arrangements with no patterns of clustering or homogeneity.

Band Combinations and Classifications

Finally, lacunarity algorithms, based on both skipping and overlapping boxes, were applied to transform bands 2, 3, and 4 of the IKONOS image. In addition, a number of layer stacks involving both lacunarity-transformed bands and original bands were generated in the following combinations:

- three original bands,
- three lacunarity-transformed bands,
- combination of three original bands and three lacunarity-transformed bands,
- combination of three original and one lacunarity-transformed band, and
- combination of two original bands and two lacunarity-transformed bands.

All window sizes were applied to stacks of three original bands and three lacunarity-transformed bands, as well as the combination of three original bands and one lacunarity-transformed band. Regarding the three separate lacunarity-transformed bands, only the 27 x 27 window was used, primarily to observe whether satisfactory accuracy levels were attained without the original bands. In the case of the combination of the three original bands and one lacunarity-transformed band, the lacunarity-transformed band 3 was used to test whether one lacunarity-transformed band could improve the accuracy of the three original bands.

Experiments were conducted with all window sizes using lacunarity-transformed band 3, as well as testing lacunarity-transformed band 4 with the original bands using the 27 x 27 window for significant differences between lacunarity-transformed band 3 and lacunarity-transformed band 4. In the case of the combination of two original bands and two lacunarity-transformed bands, band 3 and band 4 were tested using the 27 x 27 window. All code for the modified lacunarity approaches was written in the C++ programming language.

Reference Map for Error Matrix

When examining the effectiveness of different classification algorithms, "wall-to-wall" comparisons were used by checking every pixel in the image. The selected area covered 274,458 pixels (447 x 614) representing all three urban classes: grassland, commercial, and residential. Reference classes were collected by manual interpretation, using sound local area knowledge and a thorough ground survey, and then digitally delineated with negligible positional error. Overall accuracies produced by the combination of three original bands and three lacunarity-transformed bands using the overlapping box algorithm based on the 9 x 9, 15 x 15, 21 x 21, 27 x 27, 33 x 33, and 39 x 39 window sizes were 59.36%, 61.72%, 61.86%, 61.01%, 59.94%, and 61.65%, respectively. They demonstrate that overall accuracy increases slightly with expanding window sizes, despite a slight decline when using the 33 x 33 window, and it is fairly reasonable to suggest that the 15 x 15 and 21 x 21 window sizes are effective for this analysis.

However, overall accuracies produced by the combination of three original bands and three lacunarity-transformed bands using the skipping box algorithm based on the 9 x 9, 15 x 15, 21 x 21, 27 x 27, 33 x 33, and 39 x 39 window sizes were 58.31%, 59.29%, 57.40%, 56.41%, 56.92%, and 58.31%, respectively. Different window sizes produced inconsistent overall accuracies, and hence it is difficult to determine the optimal window size. In comparison, the overlapping box algorithm outperforms the skipping box algorithm, and because the overall accuracy of the original bands is 58.38%, it can be concluded that lacunarity using the overlapping box algorithm improved the classification accuracy of the original bands. However, we feel that accuracy was not significantly improved.

Overall accuracies produced by the combination of three original bands and one lacunarity-transformed band (band 3) using the overlapping box algorithm based on the 9 x 9, 15 x 15, 21 x 21, 27 x 27, 33 x 33, and 39 x 39 window sizes were 58.49%, 60.74%, 60.94%, 60.43%, 60.29%, and 61.35%, respectively. From this, we can deduce that there is only a slight difference between adding the one lacunarity-transformed band (band 3) to the original bands and adding three lacunarity-transformed bands to the original bands. However, it was also found that the combination of three original bands and three lacunarity-transformed bands produced higher accuracies.

Overall accuracies generated by the combination of three original bands and one lacunarity-transformed band (band 4) using the overlapping box algorithm for the 15 x 15 and 27 x 27 window sizes (58.14% and 58.99%, respectively) were lower than the combination with lacunarity-transformed band 3. Overall accuracies produced by the combination of three original bands and one lacunarity-transformed band (band 3) for the skipping box algorithm using the 9 x 9, 15 x 15, 21 x 21, 27 x 27, 33 x 33, and 39 x 39 window sizes were 58.87%, 60.07%, 58.60%, 57.37%, 56.93%, and 57.69%, respectively. The results show that there is no difference between adding three lacunarity-transformed bands to the original bands or adding one lacunarity-transformed band (band 3) to the original bands.

This may be mainly because the skipping box algorithm is not particularly effective in identifying texture features from image data. As mentioned earlier, we also tested the lacunarity-transformed bands generated by the overlapping box approach, individually.

These texture bands on their own gave slightly lower accuracy (57.80%) than the individual original bands, which tends to imply that spatial information alone may not be effective in identifying land-use/land-cover classes (an area we intend to research further). We also examined the combination of two original bands (bands 3 and 4) and two lacunarity-transformed bands (again bands 3 and 4) using the overlapping box algorithm based on the 27 x 27 local window. The results from this approach produced an accuracy level as high as the combination of three original bands and three lacunarity-transformed bands. This is probably because IKONOS bands 2 and 3 are visible bands and therefore highly correlated.

Urban remote sensing is currently experiencing a paradigm shift away from spectral-only classification and toward the identification of urban objects using spatial metrics and neighbourhood-based pattern recognition. Most of the conceptual restructuring is underscored by the availability of very high spatial resolution imagery, but also by the practical necessity of generating base maps of urban land use. Our work reported in this chapter is a contribution to research on spatial and textural identification of urban objects. We explored the utility of lacunarity to measure urban heterogeneity where results suggest that the overlapping box approach is more effective than the skipping box alternative. We documented that there is no significant difference between window sizes, except for the case of the 9 x 9 window, and it might be reasonable to conclude that the 15 x 15, 21 x 21, and 27 x 27 windows are the most effective sizes in our study. It was also found that the combination of two original bands (bands 3 and 4) and two lacunarity-transformed bands (bands 3 and 4) is as accurate as the combination of three original bands and three lacunarity-transformed bands.

The combination of three original bands and three lacunarity-transformed bands is only a slight improvement on the combination of three original bands and one lacunarity-transformed band (band 3). However, what is certain is that the original bands alone or the purely lacunarity-transformed bands are not effective for this type of land-use and land-cover mapping.

Although variable, our results are very much in line with other documented work on textural and spatial characterization of high spatial resolution sensor data of urban land cover. It seems metrics, be they indices of dispersion, contagion, fractal, or lacunarity, are highly sensitive to the initial image segmentation, and inextricably affected by site, time, and scene.

Another reason for small and variable improvements in urban representation is the conceptual gulf between the computational limitations of the discretized remote sensor data model and the heterogeneous and dynamic nature of the urban landscape.

Such a gulf may prove difficult to bridge immediately with either spectral or textural/spatial indices, and a more integrative approach may ultimately be more appropriate. However, for now our work on lacunarity demonstrates another contribution to the potential for spatial metrics to characterize the increasingly finer spatial resolution of remote sensor data. We hold firm to our belief that highly detailed, spatially heterogeneous urban land cover can be measured by algorithms that are equally sensitive to geometric fluctuations.

APPLICABLE TO REMOTE SENSING

The Daubert standards reviewed above will only be applied to remotely-sensed data presented through expert testimony. Remote sensing evidence will be subject to several FRE, which are applicable whether or not an expert is called to testify.

Relevancy, Authentication, and Foundation Any evidence, scientific or otherwise, must be found relevant to the case, meaning that it must make a consequential fact more or less probable than would be deemed otherwise. If used to aid witness testimony, the map must help the trier of fact understand the testimony. Once evidence is found to be relevant, it must be authenticated. Extrinsic authentication is necessary unless the map fulfills one of the self-authentication exceptions listed in Rule 902 of the FRE. A map published by the government, for instance, is self-authenticating under Rule 902(5).

Finally, the evidence must have an adequate foundation; it must be accurate and reliable. If accuracy cannot be confirmed, courts will not admit the evidence. Of these provisions, the main evidentiary hurdle for digital maps is reliability. Courts will ask where the information in the map originated, how the information was transformed into digital form, and how the map itself was created. Since computers create digital maps, the maps will face reliability challenges as computer evidence. Courts, for instance, will inquire into "computer programming errors, equipment malfunction, data entry errors, and the volume of electronic data."

Courts will also closely consider the authenticity of digital maps, particularly where the map does not meet one of the aforementioned self-authentication exceptions. As such, courts will follow Rule 901(a), requiring proof that the evidence is what its proponent claims it to be. According to Rule 901(b)(9), parties must prove that evidence encompassing a process or system, such as maps depicting remotely-sensed data, must produce an accurate result. To satisfy these rules, the experts who collected the remotely-sensed data should describe how the process operates and their involvement. Experts should also reference the data to ground information ('ground-truthing'), aerial photographs, and other maps. Logs and records of the progression from collection to presentation of the data would also verify authenticity. Technologies including steganography and cyclic redundant checksum are continually being developed to assist in ensuring the authenticity of digital imagery.

Hearsay Issues

If a map, chart, or other media is admitted to make an assertion, the evidence may be objected to on hearsay grounds. For example, remotely-sensed data could be used to create a map depicting high levels of pollution in a stream adjacent to the defendant's property. If the map is admitted to assert that the defendant caused such pollution, it may meet with a hearsay objection. If the evidence is found to be hearsay, it will only be admissible if it can be categorized as an exception to the hearsay rules. For example, Rules 803(6) and 803(8) will allow the admission of hearsay evidence that was generated by computer for use as a business or public record.

Data Characterization

A final set of rules that may pertain to the use of remotely-sensed data involve the presentation of the evidence in the courtroom. Rule 1006 allows the admission of charts, summaries, and calculations that depict a body of data too voluminous to itself be admitted into evidence for practical reasons. To avoid potential problems with admission under this rule, experts should testify that the data was correctly translated into these summary forms. If the evidence is admitted without the verification of expert testimony, Rule 1002 requires that the underlying data be admissible. For example, if a chart includes data derived from satellite photos, courts or opposing attorneys could bar the admission of the chart if the original photos do not also meet the standards of admissibility.

Constitutional Hurdles

Besides Daubert and the FRE, the Constitution presents another obstacle that remote sensing data must overcome for admission into federal courts. The main constitutional issues facing remote sensing data are allegations of invasions of privacy and warrantless searches. The Fourth Amendment states that "the right of the people to be secure in their persons, houses, papers, and effects, against unreasonable searches and seizures, shall not be violated." Two Supreme Court cases, Dow Chemical Co. v. United States and Kyllo v. United States, address the application of the Constitution to remote sensing data.

In Dow Chemical Co., the Court held that enhanced aerial photographs of an industrial facility taken by the EPA were admissible under the Fourth Amendment. The Court found that though commercial areas receive constitutional privacy protection, this protection does not extend to the outdoor areas of industrial complexes.

The Court also found that homes and their outside areas receive a higher level of protection than commercial areas. Still, in dicta the Court stated, "surveillance of private property by using highly sophisticated surveillance equipment not generally available to the public, such as satellite technology, might be constitutionally proscribed absent a warrant." The Court feared that technology providing information not available to the naked eye would reveal intimate details, for example, imaging that could reveal actions occurring inside a building (e.g., conversations behind closed doors or people transporting documents). Despite this concern, the Court noted that photos enhancing human vision were still admissible, provided that they do not reveal such intimate details.

The Supreme Court's latest decision regarding remote sensing data's privacy and search issues is Kyllo v. United States. Kyllo involved a police officer who used a thermal imaging device to detect heat emissions from a suspect's home. Declaring this search unconstitutional, the Court held that when "the Government uses a device that is not in general public use, to explore details of the home that would previously have been unknowable without physical intrusion, the surveillance" is unconstitutional. As in Dow Chemical Co., the Court emphasized that homes receive a high level of privacy protection under the Constitution. The Court held that, in the home, "all details are intimate details," strongly indicating that any information obtained by remote sensing data from a home's interior without a warrant would be inadmissible.

The Court did not define "general use" technology in either Dow Chemical Co. or Kyllo. Lower courts are left to speculate on what level of use might rise to this standard. For example, remote sensors that track wetland deterioration might be deemed "general use" technology if they are routinely used by the government, or if the public accepted their use. But if the device determined that someone illegally filled in a wetland in his or her backyard, that information could be inadmissible. The main lesson that can clearly be drawn from Dow Chemical Co. and Kyllo is that, in the absence of a warrant, remote sensing data will only gain courtroom admission if it does not include intimate details of commercial activity or any details from private homes.

REMOTE SENSING PROCESSING

During this first step the quality and consistency of the remote sensing dataset were improved. This is a crucial step in the workflow, as the structural data used for the surface reconstruction were mostly acquired from geological traces mapped over the satellite image. The accuracy in the positioning of the various horizons affects the orientation values as well as the lineament interpretation on the map. Consequently, precise geo-referencing and orthorectification are crucial to this step of the workflow. To achieve this, the remote sensing data were orthorectified using Toutin's rigorous model (Toutin et al. 2002) involving the Spot DEM and geodetic points from cartography.

The relative positioning between Spot and Aster images was evaluated by comparing 35 check points homogeneously distributed on each image, selected on characteristic pixels and at different elevations (in a range between 200 and 1600 m).

All the check points present a divergence <18 m, a value close to the Aster pixel resolution and suitable for a reliable overlapping. Successively, through multispectral analysis we improved the lithological differentiation of the images allowing a clear correlation of colours with the main sedimentary formations. Aster images have been processed with the Optimum Index Factor (OIF) technique, described by Chavez (1982), to achieve maximum discrimination between the outcropping lithologies. Pickup et al. (1993) and Ray & Murray (1994) showed the superior efficiency of this technique in desert areas over other methods.

The OIF is a statistical approach that ranks all possible threeband combinations and determines that with the maximum extractable in the false colour image. The three-band combination with the largest OIF value is expected to have high total variance within bands and low correlation coefficient between bands. Only bands from 1 to 9 (visible to short IR) were used in this statistical analysis to preserve the spatial resolution of the final image at less than 30 m per pixel. It was found that bands 3, 4 in the visible and near IR (VNIR sensor, 15 m) and 6 in the shortwave IR (SWIR sensor, 30 m) are characterized by high values of OIF. The result of the low correlation between bands is clearly reflected in the strong colour discrimination associated with the outcropping lithologies. To increase the spatial resolution of the resulting image whilst preserving the spectral information of the multispectral data, the image was pansharpened with the 2.5 m resolution panchromatic Spot image. The result is a high-resolution image with a good lithological colour differentiation. Once the remote sensing data had been processed, the images were draped over the DEM to generate the geographic information system (GIS) environment required to visualize and process the structural dataset in three dimensions. The study area was then subdivided into 33 clipped square areas (10 km each side) to reduce the computing workload of the 3D images during the mapping phase.

THE 3D MAPPING AND SURFACE RECONSTRUCTION

The traces (polylines) of the selected geological horizons have seen mapped on the final 3D image. The presence of structural relief allows the best-fitting plane of the digitized points along the bedding traces to be calculated and the associated new dip data value to be extracted. The software (developed by Fernández et al. 2004) allows the quality of the interpolated surface to be determined and this can be compared with the digitized traces, rejecting traces that could give rise to misleading values. Once the new dataset is obtained we can measure the stratigraphic separation between digitized elements that preserve azimuth and dip values. Once the stratigraphic separation vector field between each dip data point and the horizon to be reconstructed has been obtained, data can be projected accordingly.

Dip data are grouped into dip domains using a threshold of ±57deg; in azimuth and ±3° in dip. A mean plane is calculated for each dip domain and projected to the appropriate stratigraphic position to reconstruct the selected horizon. In the reconstruction of a regional-scale structure the areas where the gentle topographic relief does not allow satisfactory 3D mapping have to be considered. It is not possible to obtain extra dip data in these areas and further geological constraints are required to define the 3D model.

Further geological constraints have been added on the basis of the definition of the best-fit geometrical-structural model to account for all the available data. Such a structural model has been defined by a systematic structural analysis of the dataset to divide the structure into as many simple geometrical domains (i.e. cylindrical or conical models) as necessary. This approach would be consistent with the developable surface ideas for the 3D reconstruction of geological structures (Mynatt et al. 2007). In the proposed method we have followed the equivalent dip domain method described by Carrera et al. (2009). This method is based on the definition of the vector field grain of plunge lines associated with the cylindrical or conical domains into which any folding structure can be decomposed. Thus, if a cylindrical best fit is obtained for a dip dataset in a specific sector of the structure, a cylindrical domain and its related plunge lines can be defined. Subsequently, any data point or eventually an entire dip domain as defined with the available data can be projected along the plunge lines to fill gaps of data. In a general case of a structure characterized by multiple cylindrical domains, data will be projected across the boundary planes of the adjacent cylindrical domains, which will be the bisector planes between the planes normal to the plunge lines (characteristic surfaces; Carrera et al. 2009) to preserve bedding thickness. This method made it possible to increase data in areas where 3D mapping was not fruitful. The next step is the manual editing of the dip domain planes at different scales, considering the bedding traces as constraints.

This step improves the 3D model, establishing the validity of each dip domain by observing the relative fit with the bedding trace. It has been assumed that a dip domain is suitable for the 3D reconstruction only if the stratigraphic distance between the bedding trace and the dip domain plane is less than 15 m. If this fit has not been observed, the geometry of the surface is reshaped to match the bedding trace. This is done by removing the portions of the reconstructed surface that depart from the bedding trace by more than 15 m, then connecting the preserved portions using the Discrete Smooth Interpolator. As a result of the above-described procedure we obtain a set of planes oriented in 3D space with a value of azimuth and dip, representing the structure's frame. To interpolate the surface between the obtained planes we have used the Discrete Smooth Interpolator (Mallet 1992) implemented in the gOcad1 software. It uses a triangular mesh with an arbitrary set of nodes fixed by the user (constraints) and assigns coordinates to the other mesh nodes from multiple iterations, progressively reducing the sharpness of the final surface.

CASE STUDY: THE ANARAN ANTICLINE

Geological Setting

The Anaran anticline is located along the mountain front of the Simple Folded Belt structural unit, which represents the external part of the Zagros fold and thrust belt, in the Lurestan province. The Simple Folded Belt is characterized by NW-SEtrending giant anticlines, spectacularly folding a Palaeozoic to Cenozoic sedimentary sequence of up to 12-14 km thickness. These structures have been classically interpreted as fault propagation and detachment folds.

The key points in understanding the Simple Folded Belt structural style are the influence of basal and intermediate detachment horizons and that of pre-existing structures in the Precambrian basement (Bahroudi & Talbot 2003). The Hormuz Formation evaporites have been traditionally interpreted as the main detachment level in the Zagros fold and thrust belt, detaching the Palaeozoic-Mezosoic cover from the Precambrian basement. The detached cover has also multiple internal detachments producing structural complications such as salt migration, buttressing and disharmonic folding.

The general NW-SE trend of the Zagros is inherited from a pre-existing structural grain in the basement and can also be observed in the basement structures of the Najd fault system exposed in Nubia and Arabia.

The Mountain Front Fault, one of the key features in the study area, has also been interpreted as a result of the reactivation of basement structures in the Zagros fold and thrust belt. The anticline's axis at the southeastern termination of the Simple Folded Belt plunges into the Dezful embayment with a regional vertical drop of about 3 km. At the boundary between the Lurestan and Dezful provinces the fold swings from a NW-SE trend to an east-west trend. This east-west trend is thought to relate to the underlying BalaRud fault system. Its structural interpretation remains controversial.

On a regional scale, various models have been proposed to explain the existence of salients and embayments along the Zagros fold and thrust belt, but on the small scale the role of previous structures in controlling the location, orientation and geometry of the overlying structures is still poorly understood. A realistic 3D model of the Anaran anticline is necessary to better constrain the influence of these pre-existing structures and detachments in the structural segmentation and geometrical complexity of the area.

An initial inspection of the Anaran anticline on the satellite images reveals its complex geometry, which makes it interesting for a 3D reconstruction approach. The anticline is 100 km long, and presents strong variations in axial surface orientation and deformation style along and across strike, as well as in the stratigraphic thickness of the units involved. Because of the structural variability of this anticline, the structure was subdivided into four sub-domains for a more efficient management of the dataset. These sub-domains are: (1) NW dome: NW-SE-trending dome-like anticline closure; (2) oblique anticline: affected by north-south normal faults; (3) central anticline: emergent mountain front thrust (MFT) and affected by longitudinal NW-SE normal faults; (4) SE closure: NW-SEstriking axis.

Stratigraphy

A correlation between the main formations and related lithologies (from Alavi 2004) and the resulting colours derived from the remote sensing data.

Pre-growth Strata

The oldest rocks in the pre-folding sequence to crop out in the study area belong to the Sarvak Formation and are Late Albian to Late Turonian in age. They consist of up to 800 m of resistant, cliff-forming, grey, shallow-marine limestones (in part argillaceous and micritic and in part sparitic), containing a 30-60 m interval of greenish grey shale. The Santonian Ilam light grey shallow-marine limestones, 150 m thick, overlie the lower Turonian Sarvak strata with a sharp weathered contact, often observable in the Aster images.

The Sarvak and Ilam Formations form the upper part of the Bangestan Group, included in the carbonate platform that developed since Triassic time. The Bangestan group is overlain by c. 200 m of the Upper Cretaceous (Santonian to Mastrichtian) Gurpi Formation, a sequence of thin-bedded deep-marine rocks formed by pelagic marls (and marly limestones) and dark bluish grey hemipelagic claystones. The increasing presence of clay minerals is highlighted by the absorption peak of the Al-OH and Mg-OH minerals in the Aster short IR bands (band 6 in the generated false colour image).

The top of the Upper Cretaceous sequence is defined by a subaerial unconformity. The Palaeocene to Eocene Pabdeh Formation (400-430 m thick), with a thin purple, sandy to silty shale layer at the base, overlies the top of the weathered phosphoriterich Gurpi Formation, indicating a significant erosional unconformity. This purple shale member is clearly detected in the Aster images because of the spectral contrast between the phosphate-bearing minerals and clay minerals, and represents a guide horizon along the anticline limbs.

A c. 40m of cliff-forming chert limestone overlies the purple member, followed by thin-to medium-bedded bluish grey shale and marl and interlayers of argillaceous limestones. The Oligocene is represented by the Kalhur Member, which consists of 180 m of evaporites (gypsum, anhydrite), interbedded with grey calcareous shale, argillaceous limestone that has a very characteristic spectral sign of strong reflectance values in the visible wavelengths and high adsorption in the short IR.

The Asmari Formation is Late Oligocene to earliest Miocene in age and consists of 150 m of medium-bedded to thick-bedded limestones shoaling upward above a thin basal conglomerate ranging from fine-grained (low-energy) deep-marine marly limestone to high-energy shallow-marine skeletal grainstone interpreted as transgressive-regressive foredeep facies of the foreland basin. As a result, the rock is a highly prolific reservoir. The top of the Asmari Formation forms the flanks of the anticlines and this structural surface is overlain by the Gachsaran evaporites, deposited during the early Miocene. The Gachsaran Formation is characterized by multiple sequences of variable thickness and lithology including alternations of grey evaporites (gypsum, anhydrite, and subordinate halite), dark bituminous to red shale, and grey to red marly cross-bedded quartzose sandstone. A thick evaporite layer near its base serves as the seal for hydrocarbon reservoirs in the underlying Asmari units.

Growth Strata

According to the magnetostratigraphic dating of Homke et al. (2004), the timing of the fold-related deformation began in the Late Miocene (8.1-7.2 Ma) during the deposition of the Agha Jari Formation, a thick (up to 3000 m) succession composed of carbonate-clastic and polymictic conglomerates, calcarenites, grey cross-bedded sandstones (where chert grains are a major constituent), siltstones, marls, and lime-mudstones with interlayered lenses of gypsums. The coarse-grained Bakhtiari fluviatile conglomerates and their partly time-equivalent finer-grained Lahbari calcareous argillites and gypsum interfingers overlie the Agha Jari Formation. Their upper limit is defined by the present-day erosional level.

The 3D Reconstruction

The 3D reconstruction of the top of the Anaran anticline has been based on the reconstruction of the surfaces of the Ilam and Asmari Formations. These horizons have been selected as the reference horizons for the 3D surface reconstruction for many reasons. The Ilam Formation crops out widely in the core of the anticline, often preserving the geometry of the anticline hinge zone. The Ilam and Sarvak Formations (clearly discernible by spectral contrast) are dominantly composed of resistant, cliff-forming limestones, which make it possible to calculate the stratigraphic thickness through 3D mapping, especially along fault scarps.

The Asmari Formation limestones produce prominent structural surfaces along the flanks of the structure, which are particularly useful for establishing the frontal and back-limb geometry of the anticline. In addition, the mechanical contrast between the basal anhydrites (about 200 m) of the Gascharan Formation and the top of the Asmari Formation makes the 3D mapping of this surface easier. Other horizons have been reconstructed in some areas to better constrain the reconstruction of the reference horizons.

The digitized data include more than 1282 map bedding traces, totalling about 1000 km in length. From these traces 1180 dip values have been extracted. The resulting dip dataset has been grouped into 231 dip domains. As expected, because of the dependence of the 3D mapping quality on many geomorphological factors, the resulting dataset has a scattered distribution. Areas with considerable lack of dip data are directly related to outcrops of recessive rocks and areas covered with modern deposits. This problem was mainly present in the hanging wall of the grabens located along the anticline crest, especially in the central part, in the southeastern closure and in the frontal limb along the central and northern sectors. The increase in density of the dataset in these areas has been achieved through plunge projection. Ten sectors with a subcylindrical geometry have been identified.

No plunge projection of data was necessary in the NW dome, where the Ilam Formation generally preserves the structural surface. In contrast, along the oblique sector and in the periclinal terminations, the bedding data regression generated a large variability, demonstrating that the fold geometries cannot be easily treated as a developable surface. In these areas the 3D reconstruction has been carried out following only the dip domain method. The geometries at depth as well as in the eroded areas, where it was not possible to control the stratigraphical thickness, have been reconstructed by assuming layer-thickness preservation. Accordingly, we assume that during folding rock volumes bounded by major faults deform by flexural slip folding.

This simplification allows us to simplify the mathematical handling and to use surface and near-surface fold geometrical attributes (e.g. plunge lines, fold axial trend) to constrain the deep geometries. A good example is along the frontal limb, where the stratigraphical projection of the top of the Gachsaran anhydrites has been performed to model the buried part of the frontal limb of the Ilam Formation.

Modelling the faults that crop out in the study area has been one of the more challenging aspects of the 3D reconstruction. The main criteria used to estimate the presence of faults and to deduce the fault surface geometry from its morphological expression on the DEM are the measure of the stratigraphic throw and the mapping of the corresponding trace. Exceptionally welldeveloped fault scarps linked to the main normal faults and grabens along the central and oblique sectors make it possible to estimate the fault plane orientations.

Consequently, the reconstructed faults are the ones that considerably displace the reference horizons. A total of 64 fault planes that affect the Ilam horizon beyond c. 60 m of stratigraphic throw were identified. Less significant faults were neglected (also owing to the DEM resolution limit). The reconstruction of fault geometry at depth is difficult because of the lack of subsurface constraints. Consequently, surface outcrops of fault surfaces are used to make a qualitative interpretation of the likeliest fault geometry at depth.

STRUCTURAL IMPLICATIONS FOR THE ANARAN ANTICLINE

The 3D reconstruction of the key surfaces along the anticline has allowed for fine details of the structure to be deciphered, and their possible relationships with the fault geometry at depth to be deduced. The reconstructed geometry of the selected surfaces shows the existence of different structural patterns along the four defined sectors of the anticline. The eastern sector of the Anaran anticline can be considered as a separate double-plunging anticline of less structural complexity than the central sector. In this eastern sector the anticline displays a sigmoid shape with the two terminations striking approximately east-west. The structure, like that of the rest of the anticline, is SW-verging. In the frontal limb the bedding dip increases progressively from subhorizontal (in the hinge zone) to nearly vertical towards the foreland.

Along the frontal limb an anomalous thickness of the Gachsaran Formation can be observed. This consists of 4 km of outcropping, folded mobile salt piled up between the base of the Agaha Jari Formation and the top of the Asmari Formation. The backlimb dips about 208 northwards and has a very abrupt junction with a very steeply dipping panel close to the fold crest. This short subvertical panel is present only in this sector of the structure and has been interpreted as related to a back-thrust, which has been detached into the Jurassic evaporites, probably caused by the buttressing effect induced by the evaporites piled in front of the structure preventing the later propagation of the main detachment to the south.

The origin of this heavily folded Gachsaran sequence, located just in front of the SE sector of the anticline, may be the lateral migration of evaporites caused by the increase in displacement of the central sector towards the SW. The constant altitude of the anticline crest along this sector suggests a constant depth to the detachment for the fold-related thrust and the back-thrust in the subsurface.

The transition between the southeastern and central sector is marked by a saddle-like geometry between the two closures. At their junction, both folds strike roughly east-west. Proceeding along the central sector westwards the fold axis gradually changes its orientation from c. 095/2 to c. 120/5, reaching a higher structural level, before being sharply cut obliquely by the north-south-trending normal faults in its western termination. In the central sector the crest and the frontal limb are affected by grabens that follow the same strike as the anticline axis. One set of high-angle normal faults that reaches a vertical throw of up to 600 m is located along the hinge zone. Another set of low-angle normal faults cuts the frontal limb. They have been interpreted (Emami et al. 2010) as forming at a high angle and subsequently being rotated during the development of the anticline.

This significant expression of extensional deformation could be explained by the outer-arc extension generated during the folding of the thick deformed series, associated with the maximum curvature of the anticline. However, no relation has been observed between the areas with high curvature and the presence of normal faults anywhere along the anticline. Alternatively, considering normal faults to be generally linked with higher structural levels, it is reasonable to interpret them as being generated by the gravity collapse of the anticline crest. The 3D model provides evidence of a thrust along the frontal limb of this anticline section close to the oblique anticline section featuring north-south-trending faults.

The axis trend suddenly changes westward of the central sector coinciding with the existence of the north-south normal faults. Generally, these faults dip steeply to the west with a left-stepping en échelon organization. They show a progressive increase of the normal slip displacement from the back limb to the frontal limb. The extension in this oblique zone between the central sector and the NW dome lowered the structural level of Ilam Formation, to c. 1000 m. The normal fault displacement locally affects the general NW-SE trend of the anticline axis, generating north-south grabens that present a progressive southwards displacement increase. A second family of subvertical NE-SW-trending faults affects the frontal limb of this anticline section. They show a left lateral oblique slip component that displaces the first-order normal faults. This latter fault set probably resulted from the accommodation of differential uplift of the anticline.

In the NW dome sector the Ilam Formation preserves the dome-like structure shape of the anticline closure. It verges strongly southwestwards, with the fold strike progressively curving from SE-NW to east-west and plunging westwards along the closure. Moving from the oblique sector into the NW sector of the structure, the anticline regains the same structural level as the central sector. In addition, the aspect ratio of the fold (Burberry & Cosgrove 2010) is also comparable, indicating that both are probably detached at the same depth.

Looking at the 3D model as a whole, it can be seen that the two main sets of normal faults strike parallel and at about 458 to the local fold axis, in the central and oblique sectors respectively. In the oblique fold sector, the north-south normal faults display an en échelon geometry, with the northern fault tips aligned parallel to the NW-SE fold axis. The relation between the geometry of the normal faults and the fold axis demonstrates the simultaneous development of the anticline with both extensional systems. These two sets are, respectively, perpendicular and oblique to the shortening direction, and we interpret the development of normal faults along the oblique sector as related to an oblique ramp of a thrust at depth with a right lateral strike-slip component.

The results of this study demonstrate that the proposed approach is a powerful tool for the 3D reconstruction of geological surfaces when working with remote sensing data in inaccessible areas. It is especially useful in semi-arid regions, where the structure strongly controls the topography. It is also valid for preliminary studies previous to field work, as well as for all the stages of exploration studies on the integration of surface and subsurface data into a coherent structural model.

The processing carried out on the remote sensing images allowed the extraction of lithological information at the outcrop scale, making the mapping of the main stratigraphical formations and members possible to a level of detail higher than that of the published geological maps. The combination between the 3D dip-domain reconstruction approach and the DSI automated interpolation proves a valid solution for obtaining 3D reconstructions of complex geological structures, making considerably easier the editing and upscaling processes.

Through the systematic structural analysis and the characterization of cylindrical domains it has been possible to project dip data along the structure and to supply dip data where 3D mapping was not fruitful. The reconstruction of the Ilam and Asmari horizons along the Anaran anticline of the case study has revealed geometries that are unlikely to have been revealed by 2D section representation.

The reconstructed model can account for the simultaneous development of the north-south-trending normal faults along the oblique sector of the fold and longitudinal grabens in the central sector. The extension in the oblique sector has been related to an oblique thrust ramp at depth with a right lateral strike-slip component.

The frontal outcropping thrust, as well as its inferred geometries at depth, has been constrained with the surface structural model of the anticline. The Anaran 3D reconstruction stands as the basis for further 3D analysis developments including 3D restoration, reservoir and fracture modelling, and curvature analysis.

URBAN TEXTURAL ANALYSIS FROM REMOTE SENSOR DATA

Despite the new generation of very high spatial resolution sensor data (IKONOS from 1999 and QuickBird from 2001), predicted improvements in classification accuracy of urban land covers (and subsequent inference of urban land use) have yet to materialize substantially. Much of the obstruction to quality information extraction is still due to the traditional limitations of classifying image data representing urban areas: the high spatial arrangement of complex urban features and how to configure multispectral responses from land cover features into organized urban land-use categories (Barr, Barnsley, and Steel 2004). When launched, the desired objective of high spatial resolution sensor data was for increased clarity of terrestrial features, especially urban objects, by reducing per-pixel spectral heterogeneity and thereby improving land cover identification.

Clarity is certainly more evident in these finer-scale data than those from preceding sensors, but paradoxically this greater level of detail is also translated into many more unique per-pixel spectral combinations. For example, the residential land-use category can now be defined from much wider spectral variations, representing minute compositional mixtures of urban land covers, such as roads, houses, grasses, trees, bare soil, shrubs, and swimming pools, each conceivably a different residential land-use category.

Following on, another limitation for improved information extraction from high spatial resolution sensor data is the reliance on techniques using traditional per-pixel spectral differentiation. To us this seems counterintuitive and we would like to see more neighbourhood-related methods, using textural and spatial parameters when dealing with fine-resolution image data.

Where traditional spectral approaches are designed to identify homogeneous features regardless of shape, textural and spatial algorithms measure both the variance within and the geometric configuration of whole urban objects, respectively.

As a contribution to the growing literature, we outline an object-based pattern recognition technique that accommodates the concept of lacunarity for characterizing the textural properties of urban land cover (and therefore inferring land use) from high spatial resolution image data. In doing so, we consolidate the utility of geometric models not only for image data but for all discrete and textural spatial representations (Zhao and Stough 2005).

Indeed, the ability to characterize the shapes of individual and groups of objects is a rapid area of research in computational geometry and at the heart of the recent developments in object-based models in many geographic information system algorithms.

Recall that remote sensor data are composed of multispectral pixel vectors that represent geographical objects and their relative configuration.

We strongly adhere to the paradigm that geometric patterns, such as lacunarity, are valuable precursors for functional processes; in our application, the texture and spatial orientation of land-cover patterns derived from remote sensor data are both forerunners for analysing land-use juxtaposition and dynamic urban processes.

Lacunarity Approach

The lacunarity of an object is the counterpart of its fractal dimension. Lacunarity methods for urban analysis, and indeed many other applications in geospatial research, have already been reported by a number of researchers.

Essentially, lacunarity is related to the spatial distribution of gap or hole sizes. For low-lacunarity measurements, all gap sizes are the same and geometric objects are deemed homogeneous; conversely, for high-lacunarity gap sizes are variable and objects are therefore heterogeneous.

In other words, the variance or texture of gap sizes within the spatial delineation of geometric objects determines the level of lacunarity. Of course, textures that are homogeneous at small scales can be quite heterogeneous at large scales, and vice versa; therefore, lacunarity can be considered a scale-dependent measure of texture. Methods for calculating the lacunarity of objects were first given, in general terms, by Mandelbrot (1983) and were later implemented by various computer algorithms.

Work by Myint and Lam (2005b) developed two modified lacunarity algorithms: the binary approach, which was first introduced by Plotnick, Gardner, and O'Neill (1993), and a gray-scale routine, initially devised by Voss (1986) and used to test the effectiveness of lacunarity on high spatial resolution sensor data. This same desire to extract urban objects from fine-scale sensor data also forms the basis of this study, where we examine modifications to a differential box counting algorithm, first formulated by Dong (2000b).

Our study will also introduce two different gliding box approaches. The first uses overlapping boxes, in which the gliding box moves to a pixel next to the previous position, and the second uses skipping boxes, in which the gliding box skips the entire coverage of the previous box before moving to the next position. As a background, and according to the gliding box algorithm proposed by Allain and Cloitre (1991), n(M,r) can be defined as the number of gliding boxes with radius rand mass M.

The computations of lacunarity values are given by worked examples where the overlapping box method is demonstrated by a 4 x 4 image or local window, while a 6 x 6 image is used to illustrate the skipping box method. The 3 x 3 gliding box used in both is the base of the cube box and is always an odd number to allow the computed value to be assigned to a central cell. A column with more than one cube box may be required to cover the maximum image intensity values by stacking cube boxes on top of each other.

5

Electromagnetic Remote Sensing Process in Agriculture

Agricultural resources data are collected y aircraft and satellite-mounted instruments, which receive reflected energy from target in some frequency of the electromagnetic spectrum. The process involved in electromagnetic remote sensing system namely, data acquisition and data analysis are outlined below and a schematic diagram of electromagnetic remote sensing process.

Data Acquisition

The data acquisition process comprises of the following distinct elements, which are necessary in agricultural survey

- Energy sources
- Propagation of energy through the atmosphere
- Energy interactions with earth's surface features
- Air borne/Space borne sensors to record the reflected energy
- Generation of sensor data in the form of pictures or digital information.

Data Analysis

The data analysis process involves examining the data using various viewing instrument to analyse pictorial data, which is called the visual interpretation technique and computer to analyse digital data, a process known as digital analysis.

Reference Data

Reference data also called ground truth area an essential part of remote sensing data processing. It is used to analyse and interpret remotely sensed data, to calibrate a sensor, and to verify information extracted from remote sensing data. The spectrum relevant to remote sensing. The most common form of remote sensing was Aerial photography, in which used the visible light section of the electromagnetic spectrum. Newer sensors can acquire data in this and other sections of the electromagnetic spectrum, such as the non-visible infrared and near infrared wavelength, as will as microwaves used for radar. Many of these sensors can acquire several section of the spectrum concurrently and may be termed multispectral scanners. The electromagnetic wavelength bands with their bands with their utility in remote sensing.

PRESENT SCENARIO

Though precision farming is very much talked about in developed countries, it is still at a very nascent stage in developing countries, including India. Space Application center, ISRO, in collaboration with Central Potato Research Institute, Shimla, has initiated a study on exploring the role remote sensing for precision farming.

The study on precision agriculture has already been initiated in India, in many research institutes. Space Application Center (ISRO), Ahmedabad has started experiment in the Central Potato Research Station farm at Jalahandhar, Punjab to study the role of remote sensing in mapping the variability with respect to space and time. M S Swaminathan Research Foundation, Chennai, in collaboration with NABARD, has adopted a village in Dindigul district of Tamil Nadu for variable rate input application. Indian Agricultural Research Institute has drawn up a plan to do precision farming experiments in the institutes' farm. Project Directorate for Cropping Systems Research (PDCSR), Modipuram and Meerut (UP) in collaboration with Central Institute of Agricultural Engineering (CIAE), Bhopal also initiated variable rate input application in different cropping systems. In coming few years precision farming may help the Indian farmers to harvest the fruits of frontier technologies without compromising the quality of land.

Prospects

Precision farming, though in many cases a proven technology is still mostly restricted to developed (American and European) countries. Except for a few (Wang, 2001), there is not much literature to show the scope of its implementation in India.

We feel that, one of the major problems is the small field size. In India more than 57.8 per cent of operational holdings have size less than 1 ha. However, in the major agricultural states of Punjab, Rajastan, Haryana and Gujarat there are more than 20 per cent of agricultural lands have operational holding size of more than 4 ha. These are individual field sizes. However, when we consider contiguous field with same crop (mostly under similar management practices) the field (rather simulated field) sizes are large. Using aerial data, has found that in Patiala district of Punjab, more than 50 per cent of contiguous field sizes are larger than 15 ha. These contiguous fields can be considered a single field for the purpose of implementation of precision farming.

There is a scope of implementing precision farming for major food-grain crops such as rice, wheat, especially in the states of Punjab and Haryana. However many horticultural crops in India, which are high profit making crops, offer wide scope for precision farming.

MISCONCEPTIONS ABOUT PRECISION AGRICULTURE

There are several mistaken preconceptions about precision agriculture.

a). Precision agriculture is a cropping rather than an agricultural concept. This is due to cropping systems, in particular broad-acre cropping, being the face and driving force of PA technology. However precision farming concepts are applicable to all agricultural sectors from animals to fisheries to forestry. In fact it might be argued that precision farming concepts are more advanced in the dairy industry where the "site" becomes an individual animal, which is recorded, traced and fed individually to optimize production. These industries are just as concerned with improved productivity and quality decreased environmental impact and better risk management as the cropping industry however precision farming concepts have yet to be applied on the same scale in these areas. For example a grazer's use of advance warning meteorological data and market predictions to estimate fodder reserves and plan livestock numbers is a form of precision farming.

b). Precision agriculture in cropping equals yield mapping. Yield mapping is a crucial step and the wealth of information farmers are able to obtain from a yield map makes them very valuable. However they are only a stepping-stone in a precision farming management system. The bigger agronomic hurdle lies in retrieving the information in the yield map and using it to improve the production system. The advance of PA adoption (usefulness) in this country is may soon be bottlenecked at this point due to the lack of decision support systems (DSS) to help agronomists and farmers understand their yield maps. Yield maps may not tell the whole story either with other data sources, e.g. crop quality and soil maps, economic indicators or weather predictions, proving further information necessary for correct agronomic interpretations.

c). Precision agriculture equals sustainable agriculture. Precision agriculture is a tool to make agriculture more sustainable however it is not the total answer. Precision farming aims at maximum production efficiency with minimum environmental impact. Currently it is the potential for improved productivity (and profitability) that is driving precision farming rather than the more serious issue of long-term sustainability. Precision farming will not fix problems such as erosion and salinity by itself although it will help to reduce the risk of these problems occurring. Sensible sustainable practices still need to be used in conjunction with precision farming.

Obstacles

There are many obstacles to adoption of precision farming in developing countries in general and India in particular. Some are common to those in other regions but the others are specific to Indian conditions are as follows.

- Culture and perceptions of the users
- Small farm size
- Lack of success stories
- Heterogeneity of cropping systems and market imperfections
- Land ownership, infrastructure and institutional constraints
- Lack of local technical expertise
- Knowledge and technical gaps
- Data availability, quality and costs.

Opportunities

Despite the many obstacles listed earlier, business opportunities for precision farming technologies including GIS, GPS, RS and yield monitor systems are immense in many developing countries. The scope for funding new hardware, software and consulting industries related to precision agriculture is gradually widening. In Japan, the market in the next 5 years is estimated at about US $ 100 billion for GIS and about US $ 50 billion for GPS and RS (Srinivasan, 2001). Punjab and Haryana states in India, where farm mechanization is more common than in others, may be the first to adopt precision farming on a large scale.

Recently, the governments of certain Asian countries initiated special efforts to promote precision farming. In Japan, the Ministry of Agriculture has allocated special funds for research on remote sensing applications of precision farming. A quasi-governmental institute "Bio-oriented Technology Research

Advancement Institute (BRAIN)" is also funding research on precision farming. In Malaysia, the Malaysian Agricultural Research and Development Institute (MARDI) is promoting research on precision farming of upland rice. In other countries, the private sector, which holds or leases a large acreage, is likely to adopt precision farming sooner than the small holders.

Precision farming is useful in many situations in developing countries. Rice, wheat, sugar beet, onion, potato and cotton among the field crops and apple, grape, tea, coffee and oil palm among horticultural crops are perhaps the most relevant. Some have a very high value per acre, making excellent cases for site-specific management. For all these crops, yield mapping is the first step to determine the precise locations of the highest and lowest yield areas of the field. Researchers at Kyoto University recently developed a two-row rice harvester for determining yields on a micro plot basis.

Precision farming can bring several benefits to the sugar beet industry in Hokkaido, where the marketing system was changed in 1986 from a quantity (fresh weight) to quality (sugar yield) basis. Heavy N fertilization, a common practice here, results in excessive N levels and a decreased sugar concentration. It is now possible to estimate sugar and amide N concentration in leaves using reflectance in visible bands and root yield using reflectance in visible and infrared bands. Incorporation of such data into a GIS along with precise positioning of non-uniform areas using GPS can be used to vary fertilizer dose within a field, thereby improving productivity. Although weighing conveyor technology has been known for some time, effective yield measurement still remains the main problem in crops such as sugar beet, onion and potato. Further, the preparation of product quality maps for these crops is as important as yield maps. In India, a few researchers in the private sector initiated studies on precision agriculture in high value crops like cotton, coffee and tea. In cotton, remote sensing coupled with GIS can assist in improved precision of insect pest management and harvesting. In Sri Lanka, researchers at the Tea Research Institute are examining precision management of soil organic carbon.

In so far as dairy farming in Asia is concerned, precision farming techniques can help in improving efficiency of methods, timing and rate of application of animal wastes leading to high application efficiency and low environmental pollution. While considering soil and climatic conditions. For instance, factors determining the risk of NO3 leaching, release of N2O through denitrification and contamination of surface and ground water by runoff can be mapped and analysed. Likewise, poorly managed areas in grass lands can be identified and the optimum period for cutting on a plot basis determined.

Nutrient stress management is another area where precision farming can help Indian farmers. Most cultivated soils in India are acidic and spatial variation in pH is high. Detecting nutrient stresses using remote sensing and combining data in a GIS can help in site-specific applications of fertilizers and soil amendments such as lime, manure, compost, gypsum and sulphur. This in turn would increase fertilizer use efficiency and reduce nutrient losses. In semi arid and arid tropics, precision technologies can help growers in scheduling irrigation more profitably by varying the timing, amounts and placement of water. For example, drip irrigation, coupled with information from remotely sensed stress conditions (e.g., canopy temperature) can increase the effective use of applied water from 60 to 95 per cent there by, reducing runoff from 23 to 1 per cent and deep percolation from 18 to 4 per cent.

Pests and diseases cause huge losses to Indian crops. If remote sensing can help in detecting small problem areas caused by pathogens, timing of applications of fungicides can be optimized. Recent studies in Japan show that pre-visual crop stress or incipient crop damage can be detected using radio-controlled aircraft and near-infrared narrow-band sensors. Like wise, airborne video data and GIS have been shown to effectively detect and map black fly infestations in citrus orchards, making it possible to achieve precision in pest control. Perennial weeds, which are usually position-specific and grow in concentrated areas, are also a major problem in developing countries. Remote sensing combined with GIS and GPS can help in site-specific weed management. Although through cost-benefit analysis has not been done yet, the possible use of precision technologies in managing the environmental side effects of farming and reducing pollution is appealing.

Strategies

Precision farming is still only a concept in many developing countries and strategic support from the public and private sectors is essential to promote its rapid adoption. Successful adoption, however, comprises at least three phases including exploration, analysis and execution. Data on crop yield, soil variables, weather and other characteristics are collected and mapped in the exploratory stage, which is important for increasing the awareness among farmers of long term benefits. The approaches to data collection and mapping must, therefore, reflect local needs and resources.

In the analysis stage, factors limiting the potential yield in various areas within a field and their interrelationships are examined using GIS-based statistical modelling. Sadler et al. (1998) showed that quantitatively important yield variation may occur over distances as short as 10m, however, only some factors such as soil structure, water status, pH, nutrient levels, weeds, pests and diseases can be controlled but not the others (soil texture, weather, topography). After determining the significance of each source of variability to profitability of a particular crop and relative importance of each controllable factor, management actions can be prioritized. It must be remembered that in some low yielding areas, the reason for poor yields may be the lack of sufficient soil nutrients in the first place. In such cases, application beyond just replenishment is necessary.

Lastly, execution phase includes variable application of inputs or cultural operations. In most developing countries of Asia. However, it is not always necessary and/or possible to use variable rate applicators. Efforts must, therefore, initially focus on limiting indiscriminate use of inputs in conventional methods. Once the economic and environmental benefits are known widely, variable rate technology would be rapidly implemented at least in high value crops. To spur adoption of precision farming methods in developing countries, pilot demonstration projects must be conducted at various growers locations by involving farmers in all stages of the project. The pilot projects must attempt to answer the grower's needs and emphasize the operational implementation of technology and complete analysis of the costs and savings involved. Documentation of pilot projects would help in examining the operational weaknesses and identification of remedial measures. The projects can be used to train innovative farmers and early adopters, expose the neighboring nonparticipating farmers to the new technologies, and show the usefulness of the technology for short and long-term management.

The role of agricultural input suppliers, extension advisors and consultants in the spread of these technologies is vital. For instance, public agencies should consider supplying free data such as remotely sensed imagery to the universities and research institutes involved in precision farming research. Also, professional societies of agronomy, agricultural informatics, and engineering must provide training guidance in the use of technologies. The involvement of inter/disciplinary teams is essential in this. Small farm size will not be a major constraint, if the technologies are available through consulting, custom and rental services.

The role of agricultural cooperatives is important in dissemination of precision farming technologies to small farmers. If precision farming is considered a series of discrete services: map generation, targeted scouting, it is possible to fit these services within the structure of a progressive agricultural cooperative in each developing country.

Changes in agricultural policies are also necessary to promote the adoption of precision farming. There are basically two policy approaches: regulatory policies and market based policies. The former refer to environmental regulations on the use of farm inputs and later refer to taxes and financial incentives aimed at encouraging growers to efficiently use farm inputs. In most developing countries the lack of penalties for pollutant generation has partly contributed to an excessive use of inputs.

Subsidies on inputs and outputs and mechanisms that prevent the price system from rationing limited resources are also common. The latter include state-guaranteed crop prices, tariffs, import quotas, export subsidies. Inputs such as water and fossil fuels are usually sold at prices that are well below the real resource cost of their use, which consists not only production costs but also includes scarcity value and costs of pollution. In such cases, the formulation of policies that reflect the real scarcity value of natural resources and penalize pollution and policies such as green payments for farmers adopting techniques that would lower environmental costs can promote the adoption of precision farming technologies.

In most developing countries, the pollution effects of agriculture have been largely ignored so far because of inability to effectively monitor such effects. The advent of precision farming, and the computerization of input and output flows, will now enable such monitoring. Higher taxes on pollution farms are often recommended, but there is strong opposition to the implementation of the polluter-pays-principle concept in most countries including India.

At the same time, some consumers in India would like to see a drastic reduction in the use of pesticides and fertilizers, and are willing to pay as much as 4 to 6 times the normal price for produce such as organic vegetables, soybean and wheat. When the price elasticity of input use is low and the input costs are only a small part of the total production expenditure, as in the case of fertilizers and pesticides. Very high taxes are required to reduce their use adequately. Given the unfeasibility of such high taxes, a hybrid policy may be implemented for controlling pollution. A tax-free quota of N can be combined with taxes on additional N use.

At the research level, many issues remain to be resolved. Although some progress has been made at Space Application Center, Ahmedabad, yield monitors for small farm conditions are yet to be developed. The development of standards for the hardware and software (image transfer formats and GPS transfer formats, map projection formats) is another issue. Crop models and decision support systems must be improved by considering local resources. Data for calibration of models must be made available to increase their accuracy and/or predictability. The ability to finance a creative information venture in agriculture will affect the speed of diffusion of precision farming technologies. Commercial banks, as well as other sources of funding, have to be educated regarding the potential of precision farming. In many developing countries, it may be worthwhile to develop programmes of subsidized credit to enable R&D activities on precision farming.

DATA FORMATS FOR DIGITAL SATELLITE IMAGERY

Digital data from the various satellite systems supplied to the user in the form of computer readable tapes or CD-ROM. As no worldwide standard for the storage and transfer of remotely sensed data has been agreed upon, though the CEOS (Committee on Earth Observation Satellites) format is becoming accepted as the standard. Digital remote sensing data are often organised using one of the three common formats used to organise image data.

For an instance an image consisting of four spectral channels, which can be visualised as four superimposed images, with corresponding pixels in one band registering exactly to those in the other bands. These common formats are:

- Band Interleaved by Pixel (BIP)
- Band Interleaved by Line (BIL)
- Band Sequential (BQ).

Digital image analysis is usually conducted using Raster data structures - each image is treated as an array of values. It offers advantages for manipulation of pixel values by image processing system, as it is easy to find and locate pixels and their values.

Disadvantages becomes apparent when one needs to represent the array of pixels as discrete patches or regions, where as Vector data structures uses polygonal patches and their boundaries as fundamental units for analysis and manipulation. Though vector format is not appropriate to for digital analysis of remotely sensed data.

TOPOGRAPHIC MAP

A topographic map is a type of map characterized by large-scale detail and quantitative representation of relief, usually using contour lines in modern mapping, but historically using a variety of methods. Traditional definitions require a topographic map to show both natural and man-made features. A topographic map is typically published as a map series, made up of two or more map sheets that combine to form the whole map. A contour line is a combination of two line segments that connect but do not intersect; these represent elevation on a topographic map.

The Canadian Centre for Topographic Information provides this definition of a topographic map:

A topographic map is a detailed and accurate graphic representation of cultural and natural features on the ground.

Other authors define topographic maps by contrasting them with another type of map; they are distinguished from smaller-scale "chorographic maps" that cover large regions, "planimetric maps" that do not show elevations, and "thematic maps" that focus on specific topics.

However, in the vernacular and day to day world, the representation of relief (contours) is popularly held to define the genre, such that even small-scale maps showing relief are commonly (and erroneously, in the technical sense) called "topographic".

The study or discipline of topography, while interested in relief, is actually a much broader field of study which takes into account all natural and man made features of terrain.

History

Topographic maps are based on topographical surveys. Performed at large scales, these surveys are called topographical in the old sense of topography, showing a variety of elevations and landforms. This is in contrast to older cadastral surveys, which primarily show property and governmental boundaries. The first multi-sheet topographic map series of an entire country, the *Carte géométrique de la France,* was completed in 1789. The Great Trigonometric Survey of India, started by the East India Company in 1802, then taken over by the British Raj after 1857 was notable as a successful effort on a larger scale and for accurately determining heights of Himalayan peaks from viewpoints over one hundred miles distant.

Topographic surveys were prepared by the military to assist in planning for battle and for defensive emplacements (thus the name and history of the United Kingdom's Ordnance Survey). As such, elevation information was of vital importance.

As they evolved, topographic map series became a national resource in modern nations in planning infrastructure and resource exploitation. In the

United States, the national map-making function which had been shared by both the Army Corps of Engineers and the Department of the Interior migrated to the newly created United States Geological Survey in 1879, where it has remained since.

1913 saw the beginning of the International Map of the World initiative endeavoring to map all of Earth's significant land areas at 1:1 million scale on about one thousand sheets covering four degrees latitude by six or more degrees latitude. Excluding boundaries each sheet was 44cm high and (depending on latitude) up to 66 cm wide. Although the project eventually foundered, it left an indexing system that remains in use.

By the 1980s centralized printing of standardized topographic maps began to be eroded by databases of coordinates that could be used on computers by moderately skilled end users to view or print maps with arbitary contents, coverage and scale. For example the Federal government of the United States' *TIGER* initiative compiled interlinked databases of federal, state and local political borders and census enumeration areas, and of roadways, railroads, and water features with support for locating street addresses within street segments.

TIGER was developed in the 1980s and used in the 1990 and subsequent decennial censuses. Digital elevation models (*DEM*) were also compiled, initially from topographic maps and stereographic interpretation of aerial photographs and then from satellite photography and radar data. Since all these were government projects funded with taxes and not classified for national security reasons, the datasets were in the public domain and freely useable without fees or licensing.

TIGER and DEM datasets greatly facilitated Geographic information systems and made the Global Positioning System much useful by providing context around locations given by the technology as coordinates. Initial applications were mostly professionalized forms such as innovative surveying instruments and agency-level GIS systems tended by experts. By the mid 1990s increasingly user-friendly resources such as online mapping in two and three dimensions, integration of GPS with mobile phones and automotive navigation systems appeared. As of 2011 the future of standardized, centrally printed topographical maps is left somewhat in doubt.

Uses

Topographic maps have multiple uses in the present day: any type of geographic planning or large-scale architecture; earth sciences and many other geographic disciplines; mining and other earth-based endeavours; and recreational uses such as hiking or, in particular, orienteering, which uses highly detailed maps in its standard requirements.

Map Conventions

The various features shown on the map are represented by conventional signs or symbols. For example, colors can be used to indicate a classification of roads. These signs are usually explained in the margin of the map, or on a separately published characteristic sheet.

Topographic maps are also commonly called *contour maps* or *topo maps*. In the United States, where the primary national series is organized by a strict 7.5 minute grid, they are often called *topo quads* or quadrangles.

Topographic maps conventionally show topography, or land contours, by means of contour lines. Contour lines are curves that connect contiguous points of the same altitude (isohypse). In other words, every point on the marked line of 100 m elevation is 100 m above mean sea level.

These maps usually show not only the contours, but also any significant streams or other bodies of water, forest cover, built-up areas or individual buildings (depending on scale), and other features and points of interest.

Today, topographic maps are prepared using photogrammetric interpretation of aerial photography, LIDAR and other Remote sensing techniques. Older topographic maps were prepared using traditional surveying instruments.

Publishers of National Topographic Map Series

Although virtually the entire terrestrial surface of Earth has been mapped at scale 1:1,000,000, medium and large-scale mapping has been accomplished intensively in some countries and much less in others. Nevertheless national mapping programs listed below are only a partial selection. Several commercial vendors supply international topographic map series.

Australia

The National Mapping Information Group of Geoscience Australia is the Australian Government's national mapping agency. It provides topographic maps and data to meet the needs of the sustainable development of the nation. The Office of Spatial Data Management provides an online free map service MapConnect. These topographic maps of scales 1:250,000 and 1:100,000 are available in printed form from the Sales Centre. 1:50,000 and 1:25,000 maps are produced in conjunction with the Department of Defence.

Austria

Austrian Maps is the government agency producing topographic maps of Austria, which are distributed by *Bundesamt für Eich-und Vermessungswesen* (BEV) in Vienna. The maps are published at scales 1:25,000 1:50,000 1:200,000 and 1:500,000. Maps can also be viewed online.

Canada

The Centre for Topographic Information produces topographic maps of Canada at scales of 1:50,000 and 1:250,000. They are known as the National Topographic System (NTS). A government proposal to discontinue publishing of all hardcopy or paper topographic maps in favor of digital-only mapping data was shelved in 2006 after intense public opposition.

China

The *State Bureau of Surveying and Cartography* compiles topographic maps at 1:25,000 and 1:50,000 scales. It is reported that these maps are accurate and attractively printed in seven colors, and that successive editions show progressive improvement in accuracy.

These large-scale maps are the basis for maps at smaller scales. Maps at scales 1:4,000,000 or smaller are exported by *Cartographic Publishing House, Beijing* while larger-scale maps are restricted as state secrets. China's topographic maps follow the international system of subdivision with 1:100,000 maps spanning 30 minutes longitude by 20 minutes latitude.

Denmark

The National Survey and Cadastre of Denmark is responsible for producing topographic and nautical geodata of Denmark, Greenland and the Faroe Islands.

Finland

The National Land Survey of Finland produces the Topographic Database (accuracy 1:5000-1:10 000) and publishes topographic maps of Finland at 1:25,000 and 1:50,000. In addition topographics maps can be viewed by using a free map service MapSite.

Germany

In principle, each federal state *(Bundesland)* is in charge of producing the official topographic maps. In fact, the maps between 1:5,000 and 1:100,000 are produced and published by the land surveying offices of each federal state, the maps between 1:200,000 and 1:1,000,000 by a federal office – the Bundesamt für Kartographie und Geodäsie (BKG) in Frankfurt am Main.

Greece

Topographic maps for general use are available at 1:50,000 and 1:100,000 from the *Hellenic Military Geographical Service (HMGS)*. They use a national projection system called EGSA'87, which is a Transverse Mercatorial Projection mapping Greece in one zone. A few areas are also available at 1:25,000. Some private firms sell topographic maps of national parks based on HMGS topography.

India

The Survey of India is responsible for all topographic control, surveys and mapping of India.

Japan

The Geographical Survey Institute of Japan is responsible for base mapping of Japan. Standard map scales are 1:25,000, 1:50,000, 1:200,000 and 1:500,000.

New Zealand

Land Information New Zealand is the government agency responsible for providing up-to-date topographic mapping. LINZ topographic maps cover all of New Zealand, offshore islands, some Pacific Islands and the Ross Sea Region. The standard issue *NZTopo* map series was published September 2009 at 1:50,000 (NZTopo50), and 1:250,000 (NZTopo250). Vector data from the New Zealand Topographic Database (NZTopo) is also available.

France

The Institut Géographique National (IGN) produces topographic maps of France at 1:25,000 and 1:50,000. In addition, topographic maps are freely accessible online, through the Géoportail website.

Pakistan

The responsibility for topographic mapping and aerial photography lies with the Surveyor General of Pakistan [SGP]. Established in 1947, the Survey of Pakistan (SOP) is based in Rawalpindi with a number of regional offices distributed at urban centers throughout Pakistan. SGP is a civil organization which, for security reasons, is headed by a Surveyor General and works under the strict control of Army General Headquarters (GHQ). Colonel C.A.K. Innes-Wilson, a Royal Engineers officer who joined the Survey of India which mapped the subcontinent, was the first Surveyor General of Pakistan. All departments which require topographic maps make their request to SGP and many are permanently registered with it for mapping and aerial photographs procurement. The SOP performs these functions under the auspices of the Ministry of Defence (MOD). Organisationally, the SOP is overseen by the Surveyor General (SG) who is a direct military appointee and a senior uniformed officer. The SG reports directly to the Secretary of Defence. Under the SG are two Deputy SG's (I and II) who manage the operational departments of the agency and a Senior Technical Advisor. These departments are divided into Regional Directorates for Topographic Mapping including the Northern region centred in Peshawar, Eastern region (Lahore), Western region (Quetta) and finally, the Southern region in Karachi. Responsibility for fields surveys and the maintenance/update of topographic maps are sub-divided according to these geographic areas.

Russia

Detailed, accurate topographic maps have long been a military priority. They are currently produced by the *Military-topographic service of armed forces of the Russian Federation*. Military topographic mapping departments held other titles in the Russian Empire since 1793 and in the Soviet Union where these maps also came to be used for internal control and economic development.

When Germany invaded in 1941, detailed maps from the USSR's western borders to the Volga River became an urgent task, accomplished in less than one year. After the war years the entire Soviet Union was mapped at scales down to 1:25,000 — even 1:10,000 for the agriculturally productive fraction. The rest of the world except Antarctica is believed to have been mapped at scales down to 1:200,000, with regions of special interest down to 1:50,000 and many urban areas to 1:10,000. In all there may have been over one million map sheets of high quality and detail. Soviet maps were also notable for their consistent global indexing system. These advantages held for Soviet military maps of other countries, although there were some errors due to faulty intelligence.

Soviet maps for domestic civilian purposes were often of lower quality. From 1919 to 1967 they were produced by *Head geodesic adminstration*, then by *Chief administration of geodesy and cartography*. Now (June 2011) civilian maps are produced by the *Federal agency for geodesy and carography*.

Soviet military maps were state secrets. After the 1991 breakup of the Soviet Union, many maps leaked into the public domain and are available for download. Map scales 1:100.000-1:500.000 can be viewed online.

Spain

The Instituto Geográfico Nacional (IGN) is responsible for the official topographic maps. It does use six scales that cover all the Spanish territory: 1:25,000, 1:50,000, 1:200,000, 1:500,000, 1:1,000,000 and 1:2,000,000. The commonest scale is the first one, which utilizes the UTM system.

South Africa

The Chief Directorate: National Geo-spatial Information (CD:NGI) produces three topographic map series, each covering the whole country, at scales 1:50 000, 1:250 000, and 1:500 000.

Switzerland

Swisstopo (the Federal Office of Topography) produces topographic maps of Switzerland at seven different scales.

United Kingdom

The Ordnance Survey (OS) produces topographic map series covering the United Kingdom at 1:25,000 and 1:50,000 scales. The 1:25,000 scale is known as the "Explorer" series, and include an "OL" (Outdoor Leisure) sub-series for areas of special interest to hikers and walkers. It replaced the "Pathfinder" series, which was less colourful and covered a smaller area on each map. The 1:50,000 scale is known as the "Landranger" and carries a distinctive pink cover. More detailed mapping as fine as 1:10000 cover some parts of the country. The 1:25K and 1:50K metric scales are easily coordinated with standard romer scales on currently available compasses and plotting tools. The Ordnance Survey maintains a mapping database from which they can print specialist maps at virtually any scale. Ordnance Survey National Grid divides the U.K. into cells 500 km. and 100 km.square on a Transverse Mercator grid aligned true North-South along the 2°W meridian. OS map products are based on this grid.

United States

The United States Geological Survey (USGS), a civilian Federal agency, produces several national series of topographic maps which vary in scale and extent, with some wide gaps in coverage, notably the complete absence of 1:50,000 scale topographic maps or their equivalent. The largest (both in terms of scale and quantity) and best-known topographic series is the 7.5-minute or 1:24,000 quadrangle. This scale is virtually unique to the United States and offers a convenient relation to customary units of measure where 1 map inch equals. 2,000 actual feet. USGS also publishes 1:25,000 maps for metric uses, 1:100,000 maps covering 30 minutes latitude by one degree longitude, 1:250,000 covering one by two degrees, and state maps at 1:500,000 with California, Michigan and Montana needing two sheets while Texas has four. Alaska is mapped on a single sheet, at scales ranging from 1:1,584,000 to 1:12,000,000.

- National Geographic TOPO offers topographic maps on CD-ROM by state. While as yet uncomplete nationwide, efforts are underway to accomplish this.

ADVANTAGES OF REMOTE SENSING TECHNIQUES IN AGRICULTURAL SURVEY

With the primary aim of improving the present means of generating agricultural data, a number of specific advantages may result form the use of remote sensing techniques. Vantage point. Because the agricultural landscape depends upon the sun as a source of energy, it is exposed to the aerial view and, consequently, is ideally suited or remote sensing techniques.

Coverage. With the use of high-altitude sensor platforms, it is now possible to record extensive areas on a single image. The advent of high-flying aircraft and satellites, single high quality images covering thousand of square miles

Permanent record. After an image is obtained, it serves as a permanent record of a landscape at a point in time which agriculture changes can be monitored and evaluated.

Mapping Base. Certain types of remote sensing imagery are, in essence, pictorial maps of the landscape and after rectification (if needed), allow for precise measurement (such as field acreages) to be made on the imagery, obviating time-consuming on the ground surveys.

These images may also aid ground data sampling by serving as a base map for location agriculture features while in the field, and also as a base for the selection of ground sampling point or areas.

Cost savings. The costs are relatively small when compared with the benefits, which can be obtained form interpretation of satellite imagery.

Real-time capability. The rapidly with which imagery can be obtained and interpreted may help to eliminate the lock of timeliness which plagues, so many agricultural survey.

Other advantages of Remote Sensing:

- Easy data acquisition over inaccessible area.
- Data acquisition at different scales and resolutions

- The images are analysed in the laboratory, thus reducing the amount of fieldwork.
- Colour composites can be produced from three individual band images, which provide better details of the area then a single band image or aerial photograph.
- Stereo-satellite data may be used for three-dimensional studies. At present, all advantages listed above have been demonstrated either operationally or experimentally:

CROPPING SYSTEMS THROUGH REMOTE SENSING

The major cropping systems are rice-wheat, rice-mustard-rice, cotton-wheat, pearl millet-wheat, rice-sugarcane and maize-wheat in Indo-Gangetic Plains of India. The Indo Gangetic plains regions are divide in four parts *viz.* Trans Gangetic Plain, Upper Gangetic Plain region, Middle Gangetic Plain and Lower Gangetic Plain region. The crops are rice, wheat, maize, cotton, sugarcane and pearl millet. These crops are grown in large areas of IGP in rabi and kharif seasons.

The cropping systems are largely affected due to pest. The major pests *viz.* aphid, stem fly, seed fly, Thrips, Shoot fly (Atherigona soccata), pod borer, moth and ball worms are affected seasonal crops. Due to affect of pest, adverse affected on the crop production and yield. Remote sensing technology plays a great role to prevent and control and management of pest. The insect affected crop and normal crops are give different tonal variation in imageries. The LISS III and LISS IV imageries are very helpful for detection of affected area. The affected areas are identified with the help of GPS (LEICA Differential GPS). The GPS give exact location and position on the imagery. The image interpretation keys shape, size, tone, texture and association help to detection to visual on the imagery and confirmation during ground survey. After image analysis found the great variability in tonal variation on the imagery.

Normal crops give red, bright red and dark red with smooth texture but pest affected area give pink, yellow and yellow pinkish red colour with irregular shape and rough texture. Remote sensing technology help to season wise analysis of pest in affected area and help to control through crop rotation, if possible and through chemically.

The different types of information on natural resources such as soil survey information, land use, vegetation information etc. can be extracted from the same satellite data. Remote sensing as a tool will give information regularly and at a low cost to enable timely reclamation. In India, production forecasting of certain crops, crop yield modeling and crop stress detection are done using remote sensing data. The first satellite remote sensed data became available in 1972. Therefore, we had the opportunity to get an overall synoptic view only after 1972. Currently a number of spacecraft imaging systems are operating using remote sensing sensors. Some of the current image systems from spacecraft platform include Indian remote sensing satellite (IRS), SPOT and IKONOS etc.

The variation between signatures, when the bands are assigned the primarily colours of red, blue and green, one can visualize the image as a colour one. Vegetation appears red in such a colour composite where NIR band is assigned as red, due to high reflectivity of vegetation in this band. This is typically termed as False Colour Composite (FCC).

The temporal IRS data were screened to select cloud free data during winter and summer season. Care was taken to select data pertaining to the period when crop coverage was with full foliage. The villages appear bright while water bodies had low backscatter and appeared in dark black colour. This was used to map the villages and water bodies using a decision rule classifier. All these two classes: villages and water bodies were combined and made a mask. The rest of the area was considered as potential agricultural area.

Detection and identification of plant diseases and planning effective control measures are important to sustain crop production. One of the potential applications of remote sensing technique in agriculture is the detection of plant disease an extensive areas before the symptoms clearly appear on the plant leaves. This has been clearly demonstrated for the first time in remote sensing studies on root wilt disease on coconut plans in Kerala under a NASA-ISRO-IARI collaborative programme (Dakshinamurti, 1969 and Dakshinamurti et al., 1970). Healthy plants give a higher reflectance in the near-infrared region and a lower one in the visible region. Diseased plants show a higher reflectance in the visible spectrum and a lower one in infrared region. This principle can be used in distinguishing healthy and diseased vegetation and assessing vegetation damage due to diseases.

Crop intensification is a viable proposition as far as India is concerned as it is endowed with favorable climatic condition for growing a diverse range of crops practically through out the year (Panigrahy and Manjunath, 2003). This concept of looking to the agricultural land as a system is known as the cropping systems approach that seeks to sustain the increase in the food production by judicious combination of crop intensification and yield improvement.

Cropping pattern is defined as the yearly sequence and spatial arrangement of crops or of crops and fallow on a given area. The cropping pattern is categorized on the basis of crop season as Kharif, rabi and summer season cropping pattern.

Growing two, three and multiple cropping patterns. The productive base of cropping system is plant growth, which is influenced by management and environment. Management here includes all those components associated with crop production. These include, crop area, crop biomass, crop rotation crop calendar, crop biomass, time and spread of sowing and harvest. Environment is the invariant resource of soil, physiography and climate.

Study Area

IGP region divided in four agro climatic regions TGP (Trans Gangetic Plain), UGP (Upper Gangetic Plain), MGP (Middle Gangetic Plain) and LGP (Lower Gangetic Plain).

These four zones are divided in 16th sub zone. The Indo Gangetic Plain region of India has total geographical area of about 47.4 million hectares (Pandey et al. 2006).

The rainfall ranges from high as 3000 to low 300 mm. Administratively, the states are divided out of 203 but only 173 districts are belong to IGP and encompasses the entire states of Punjab, Haryana, Uttar Pradesh, Bihar and West Bengal.

The major rivers of IGP are Ganga, Yamuna, Satluj and Ghagra. Soils of this region are loamy to sandy loam. The percentage of the organic carbon is enough to other region. The study was confined to Indo Gangetic Plain Region. These four agro climatic regions as defined by MOA (1997) and Agro-climatic Zones of India Planning Commission Report (2005) were further classified into 16 different sub regions. Three cropping seasons namely *viz.* Kharif (June to September), Rabi (October to February) and Zaid (from March to May) were taken into consideration.

Material and Methodology

i. IRS IC /1D LISS III (FCC) digital data

ii. District maps and state maps

iii. Indian Agro climatic zone map

iv. Global Positioning system (TOPCON and GARMIN) and

v. Performa for field data information

These data and instruments were used in the study.

The methodology consists of the study and analysis of remote sensing data, collection of ground truth information, collection of cropped and pests of field data and creation of data base for all the required physical and cultural parameters and overlying, integration and analysis of all spatial, non spatial and attribute data bases.

Methodology

Analysis of remotely – sensed data in different forms and formats was analyzed after examining the suitability of various products. The study proceeded as follows:

1. Preliminary study of the IRS IC /1D LISS III – FCC.
2. Study of false colour composites on digital (onscreen) or paper prints
3. Preparation of image interpretation keys
4. Collection of ground truth with GPS and
5. Preparation final output map

The digital remote sensing data were processed and analyzed using the ERDAS IMAGINE 8.6 software. Initially, the digital data of scenes were geo-referenced. The Kharif and rabi season images were considered as the base since it is very sharp and clear. This image is first geo referenced by taking various control points from the SOI toposhits. The projection type used is 'polyconic' with the spheroid and datum as 'modified Everest'. Subsequently, image-to-image registration was performed in order to register the second image and so on. Some sharp and easily identified features like crossing of roads, railways, canals, bridges etc. were located on both images and were selected as ground control points. Some GCP's with rms error beyond threshold were deleted and replaced by other points so as to maintain acceptable positional accuracy.

Preparation of Questionnaires

The necessity for acquiring such as information was felt on the basis of secondary information and the type of information acquired by direct observation during the reconnaissance field visit.

Preparation of Base Map

The base maps indicating permanent and artificial features such as canal, highway, roads, railway, irrigation network, and districts boundary were prepared in ARC GIS 8.3 software. For this, Survey of India sheets were used as reference materials. The permanent fixtures were transferred in digital format through digitization.

Survey of the Area/Reconnaissance

Surveyed of the area season wise in a year *viz.* kharif (June), rabi (January) and zaid (April). Reconnaissance of the area of IGP's states, collected normal and pest affected crop samples and related information along with GPS points in different fields. Global Positioning System burgeoning technology, which provides unequalled accuracy and flexibility of positioning for navigation surveying and GIS data capture. The GPS uses satellites and computers to compute position any where on the earth. GPS *viz.* TOPCON, GARMIN and Differential GPS are very good for agricultural purpose.

Collection of Ground Truth Information

Field visits were made to establish the relationship between the image elements, namely colour, texture, pattern, shape, size, association, etc. and the existence of normal crop and pest affected crop. Initially, reconnaissance traverse of the area was made to assess the trafficability and to precisely located sample areas, the pockets of land which were interpreted as normal as well as pest affected crop were precisely located on the ground with the help of topographical sheet and observations with respect to terrain conditions. Observations were, however, also made outside the sample areas randomly, in order to validate the relationship already established between image elements, and normal and pest affected crop areas. During surveyed collected crops and pests information 200 points with geo coordinate information these information overlaid on the imagery and observed visual as well as digital.

GIS ANALYSIS FOR CROPPING SYSTEMS

The districts polygon coverage were digitized for the IGP region in India covering the states of Punjab, Haryana, Uttar Pradesh, Bihar, West Bengal, and a part of Rajasthan in ARC GIS system. A database was developed for analysis of cropping systems in MS Excel, converted dBASE IV format then finally converted in.shp file to overlay analysis in ARC GIS as well as ERDAS IMAGINE.

Agro climatic zone, sub zone, districts and states information were taken from planning commission report. The coverage has been generalized based on the yield in to two classes, *i.e.* Normal crop and Pest-affected crop depending upon the range of crop survey data.

Climatic Factors and Fire Blight

A very important aspect of fire blight management involves monitoring of weather for the specific conditions that govern the build-up of inoculum in the orchard, the blossom infection process and the appearance of symptoms. A weather station record of daily minimum and maximum temperatures and rainfall amounts is needed. When 50 percent of the buds show green tissues, begin keeping a daily record of the cumulative degree days (DD) greater than 55'F (12.7 ºC).This information can be used to signal when symptoms are likely to appear in the orchard for blossom blight [103 DD greater than 55 F (57 DD greater than 12.7 ºC) after infection], canker blight [about 300 DD greater than 55 F (167 DD greater than 12.70 ºC) after green tip], and early shoot blight [about 103 DD greater than 55 F (57 DD greater than 12.7 ºC) after blossom blight or canker blight symptoms appear]

PRECISION AGRICULTURE

Precision agriculture is an agricultural system that has the potential of dramatically changing agriculture in this 21st century. Precision agriculture lends it self to most agricultural applications and can be implemented at whatever levels are required. Precision agriculture is based on information technology, which enables the producer to collect information and data for better decision making. Precision agriculture is a pro-active approach that reduces some of the risk and variables common to agriculture. Precision agriculture is more environmentally sound and is and integral part in sustaining natural resources. To better understand the need for an accurate definition of precision agriculture lets look at how precision agriculture is being considered. Precision agriculture is considered a concept, management strategy, and even a philosophy. It is said, "Precision agriculture is a phrase that captures the imagination of many concerned with the production of food, feed, and fiber." The concept of precision agriculture offers the promise of increasing productivity while decreasing production cost and minimizing environmental impacts. Precision agriculture conjures up images of farmers overcoming the elements with computerized machinery that is precisely controlled via satellites and local sensors and using planning software that accurately predicts crop development. This image has been called the future of agriculture.

Precision farming is characterised by a number of sophisticated tools that assist in monitoring variation and managing inputs. These include:

- Global Positioning System (GPS) – a referencing device capable of identifying sites within a field;
- Sensors and dataloggers – crop, soil and climate information can be monitored at a high frequency using these technologies;

- Geographic Information Systems (GIS) – maps of these attributes can be generated and analysed using simple browsers or complex models.

GPS ROLE IN PRECISION AGRICULTURE

Precision Agriculture is doing the right thing, at the right place, at the right time. Knowing the right thing to do may involve all kinds of high tech equipments and fancy statistics or other analysis. Doing the right thing however starts with good managers and good operators doing a good job of using common tools such as planters, fertilizer applicators, harvesters and whatever else might be needed..

The use of GPS in Agriculture is limited but it is fair to expect wide spread use of GPS in future. Recently a GPS-based crop duster (precision GPS Helicopter), which can spray an area as small as 4 X 4 mtr. is attracting great attention. Some progressive farmers are now beginning to use GPS for recording observations. Such as weed growth, unusual plant stress, colouring and growth conditions, which can then be mapped with a GIS programmes. In the years to come, GPS system role in precision agriculture may help the Indian farmers to harvest the fruits of frontier technologies without compromising the quality of land and produce.

STATUS OF PRECISION FARMING IN INDIA

Hence, the first thing that comes to mind is that, this system is not for developing countries, especially India, where the farmers are poor, farming is mostly subsistent and the land holding size is small. But, this is far from the truth as this approach has a large potential for improving the agricultural production in developing world. Imagine this situation where a farmer goes to his field with a GPS (Global Positioning System)-guided tractor. The GPS senses the exact location of tractor within the field. It sends signals to the computer fixed on to the tractor, which has a Geographical Information System (GIS), storing the soil nutrient requirement map in it. The GIS, in consultation with a Decision Support System would decide what is the exact requirement of fertilisers for that location. It then commands a variable rate fertiliser applicator, which is again attached with the tractor, to apply the exact dosage at the precise location of farm.

But, this is what precision farming means to large growers in the highly developed parts of the globe. To make it clearer, Precision Farming is the system of matching of resource application and agronomic practices with soil attributes and crop requirements since they vary across a field.

Tata Kisan Kendra : The concept of precision farming being implemented by the TKKs has the potential to catapult rural India from the bullock-cart age into the new era of satellites and IT. TCL's extension services, brought to farmers through the TKKs, use remote-sensing technology to analyse soil, inform about crop health, pest attacks and coverage of various crops predicting the final output. This helps farmers adapt quickly to changing conditions. The result: healthier crops, higher yields and enhanced incomes for farmers.

Government organization: Precision Agriculture models are not complete, unless the parameters related to empowerment of the farmers; especially small and marginal farmers are integrated. Now it is the turn of good news to the Indian farming community.

Some of the research institutes such as Space Applications Centre (ISRO), M.S. Swaminathan Research Foundation, Chennai, Indian Agricultural Research Institute, New Delhi, and Project Directorate of Cropping Systems Research, Modipuram, had started working in this direction and in soon it will help the Indian farmers harvest the fruits of frontier technologies without compromising on the quality of land.

According to the Exim Bank officials, though the research and development on PF is currently at a nascent stage in the country, the efforts being put on by the four research institutes were expected to turn the green revolution into an evergreen revolution.

In this context, ISRO has also initiated Gramsat project in Orissa. In the line of JDCP, the Gramsat project aims at empowering the people especially the poor and marginalized, by awareness building and access to information and services. Towards this, a network of one-way video and two-way audio Forecasting the yield of mono and multiple crops is being done at NRSA.

Acreage estimates and crop inventory is being done during Kharif and Rabi seasons for Rice, which is the major crop grown in our India. Other crops like Banana, Chillies, Cotton, Maize, Sugarcane and Tobacco are also being inventoried. Satellite data can also delineate different crops that are grownin the same area, and an inventory of each of the crops can be done.

Remote Sensing and Sensors for PF

Precision farming needs information about mean characteristics of small, relatively homogeneous management zones.

This is for Data acquisition of the farms to find the soil, vegetation and other parameters that are amenable for remote sensing. Remote sensing techniques play an important role in precision farming by providing continuous acquired data of agricultural crops. Remote sensors image vegetation, which is growing on different soil types with different water availability, substrate, impact of cultivation, and relief Sensors use for the following applications:

- Soil Properties Sensing: Soil Texture, Structure, and Physical Condition Soil Moisture; Soil Nutrients.
- Crop Sensing: Plant Population; Crop Stress and Nutrient Status.
- Yield Monitoring Systems: Crop Yield; Harvest Swath Width; Crop Moisture:
- Variable Rate Technology Systems: Fertilizer flow; Weed detection, pressure sensors

SOFTWARE SOLUTION FOR PRECISION FARMING

Linux is a good platform for doing this research-oriented work. Much of the analysis can be translated into such mainstream topics as signal processing or multi-dimensional statistics.

Some of the best software for exploring software in these topics is the product of government and university research and is "free"—an important quality in tight budgets. GRASS, xldlas and Santis are three packages which are very helpful in precision farming.

6

Soil Surveys Remote Sensing

Soil maps are also sometimes used to determine management zones. Soil maps are becoming part of the GIS database. The grid sampling technique takes separate soil samples from uniform sized grids laid out over the field. A problem with this type of sampling is the variability that can exist in soil types with in each grid. This variability makes it much tougher to determine soil characteristics within the grid for crop input management purposes. To minimize this problem smaller grids are required which then requires many more soil samples to be take for a larger number of grids. Soil samples can become a major cost of precision farming.

An alternative to grid sampling is targeted or zone sampling. The soil samples are located in homogeneous management zones instead of uniformly spaced grids. The zones are laid out using a process similar to computer based unsupervised image classification. Images obtained from multispectral remote sensors are taken of the vegetated areas of the field. The pixel digital numbers for each band are separated into statistically separable clusters that are classified into homogeneous zones. This cuts down on the soil, terrain, plant growth, and other variability within each area to be managed; thus fewer soil samples are needed for each area.

Except for county soil surveys remote sensing has not gained wide acceptance as a mapping tool for soil characteristics. This is because "the reflectance characteristics of the desired soil properties (*e.g.*, organic matter, texture, iron content) are often confused by variability in soil moisture content, surface roughness, climate factors, solar zenith angle, and view angle".

Drainage Maps

Subsurface drainage tile lines that have been installed, as long as 50 or more years ago are still partially or totally functional today. Often the existence or location of older tile lines has been lost as landowners die or sell their property. Some states, including Iowa, are now starting to require landowners to prepare and file drainage plats with county recorders when new tile lines are installed. It is desirable to have accurate drain tile maps for maintenance purposes or for the installation of new additional tile line systems. Installation of new tile lines may cut through old tile lines at unknown locations. Building livestock manure lagoons, which cut through old unknown and uncharted tile lines may cause environmental damage from manure leaking through the old tile lines.

Colour infrared ("CIR") aerial photographs have been shown to be an effective tool in locating unknown subsurface tile lines. The image data is digitized for preprocessing and then geo-referenced using ground control points. The CIR photographs show different tones of gray depending on soil type and moisture. By filtering out spectral reflectance differences due to soil type, soil moisture content in dry soils that have a higher reflectance can be identified from lower reflectance wet soils. The resulting image shows were the tile lines are located and whether they are working properly.

Normal colour aerial photographs can also be used to locate tile lines. Simple colour photographs offer tile line images similar to CIR but at a lower cost. If the soil is too dry such, the tile lines will not be visible in the image. The images similar, must be acquired when the soil is bare and within a few days after an adequate rain. High resolution and on demand temporal availability make images acquired from aircraft platforms ideal for acquiring this kind of image data.

Variable Rate Technology

One method of controlling variability within the field is VRT. VRT allows the grower to apply the quantity of crop inputs needed at a precise location in the field based on the individual characteristics of that location. Crop inputs that can be varied in their application commonly include tillage, fertilizer, weed control, insect control, plant variety, plant population, and irrigation. Typical VRT system components include a computer controller, GPS receiver, and GIS map database. The computer controller adjusts the equipment application rate of the crop input applied. The computer controller is integrated with the GIS database, which contains the flow rate instructions for the application equipment. A GPS receiver is linked to the computer. The computer controller uses the location coordinates from the GPS unit to find the equipment location on the map provided by the GIS unit. The computer controller reads the instructions from the GIS system and varies the rate of the crop input being applied as the equipment crosses the field. The computer controller will record the actual rates applied at each location in the field and store the information in the GIS system, thus maintaining precise field maps of materials applied.

Although VRT can control inputs applied to crops, it cannot control factors such as soil type, weather climate, and topography that are fixed.

Monitor Crop Health

Remote sensing data and images provide farmers with the ability to monitor the health and condition of crops. Multispectral remote sensing can detect reflected light that is not visible to the naked eye. The chlorophyll in the plant leaf reflects green light while absorbing most of the blue and red lightwaves emitted from the sun. Stressed plants reflect various wavelengths of light that are different from healthy plants. Healthy plants reflect more infrared energy from the spongy mesophyll plant leaf tissue than stressed plants. By being able to detect areas of plant stress before its becomes visible, farmers will have additional time to analyze the problem area and apply a treatment.

Water Stress

The use of remote sensors to directly measure soil moisture has had very limited success. Synthetic Aperture Radar ("SAR") sensors are sensitive to soil moisture and they have been used to directly measure soil moisture. SAR data requires extensive use of processing to remove surface induced noise such as soil surface roughness, vegetation, and topography. A crop evapotranspiration rate decrease is an indicator of crop water stress or other crop problems such as plant disease or insect infestation. Remote sensing images have been combined with a crop water stress index ("CWSI") model to measure field variations.

Simple panchromatic aerial photographs have been used to spot irrigation equipment problems. Strips in the vegetation images point to problems with water application rates from defective water nozzles.

Weed Management

One goal of precision farming is to cut crop production inputs, which result in cost and environmental savings. Conventional farming methods apply herbicides to the entire field. Site-specific variable-rate application puts the herbicide where the weeds are. Aerial remote sensing has not yet proved to be very useful in monitoring and locating dispersed weed populations. Some difficulties encountered are that weeds often will be dispersed throughout a crop that is spectrally similar, and very large-scale high resolution images will be needed for detection and identification. The use of machine vision technology systems to detect and identify weeds places remote sensors directly on the sprayer equipment. Being close to the crop allows for very high spatial resolutions. Machine vision systems have the ability to be used in the field with the real-time capabilities that are necessary to control sprayer equipment.

Insect Detection

Aerial or satellite remote sensing has not been successfully used to identify and locate insects directly. Indirect detection of insects through the detection of plant stress has generally not been used in annual crops. The economic injury level for treatment is usually exceeded by the time plant stress is detected by remote sensing. Entomologists prefer to do direct in field scouting in order to detect insects in time for chemical treatments to be effective and economical.

Nutrient Stress

Plant nitrogen stress areas can be located in the field using high-resolution colour infrared aerial images. The reflectance of near infrared, visible red and visible green wavelengths have a high correlation to the amount of applied nitrogen in the field. Canopy reflectance of red provides a good estimate of actual crop yields.

Yield Forecasting

Plant tissue absorbs much of the red light band and is very reflective of energy in near infrared ("NIR") wavebands. The ratio of these two bands is referred to as the vegetation index ("VI"). The difference of red and NIR measurements divided by their sum is normalized difference VI ("NDVI"). For crops such as grain sorghum, production yields, leaf area index ("LAI"), crop height and biomass have been correlated with NDVI data obtained from multispectral images. In order to get reasonably accurate yield predictions this data must be combined with input from weather models during the growing season.

APPLICATION OF REMOTE SENSING TECHNIQUES FOR AGRICULTURAL SURVEY

The specific application of remote sensing techniques can be used for i) detection ii) identification iii) measurement iv) monitoring of agricultural phenomena.

Area of Specific Applications

a) Applicable to crop survey

1. Crop identification
2. Crop acreage
3. Crop vigor
4. Crop density
5. Crop maturity
6. Growth rates
7. Yield forecasting
8. Actual yield
9. Soil fertility
10. Effects of fertilizes
11. Soil toxicity
12. Soil moisture
13. Water quality
14. Irrigation requirement
15. Insect infestations
16. Disease infestations
17. Water availability
18. Location of canals.

b) Applicable to range survey:

1. Delineation of forest types
2. Condition of range
3. Carrying capacity
4. Forage
5. Time of seasonal change
6. Location of water
7. Water quality
8. Soil fertility
9. Soil moisture
10. Insect infestations
11. Wildlife inventory.

c) Applicable to livestock survey:

1. Cattle population
2. Sheep population

3. Pig population
4. Poultry Population
5. Age sex distribution
6. Distribution of animals
7. Animal behavior
8. Disease identification
9. Types of farm buildings.

The use of remote sensing technology has been rapidly expanded for the development of key sectors. This paper highlights the fact that the remote sensing techniques will continue to be very important factor in the improvement of present system of acquiring agricultural data. The remote sensing provides various platforms for agricultural survey. Satellite imagery has unique ability to provide the actual synoptic vies of large area at a time, which is not possible for conventional survey methods and also the process of data acquisition and analysis are very fast through GIS (Geographic Information System) as compared to the conventional methods. The different features of agriculture are acquired by characteristic, spectral reflectance, spectral signature of agriculture and associated phenomena through EMR. In General the research paper emphasizes the utmost need of timeliness and accuracy of the output generated by remote sensing techniques and its calibration with ground-truth and other information systems like aerial photography and satellite imagery etc. Further, the importance of remote sensing with special reference to agricultural sector involving crop acreage, crop production, rangeland and livestock.

REMOTE SENSING IN AGRICULTURAL SURVEY

Remote sensing is nothing but a means to get the reliable information about an object without being in physical contact with the object. It is on the observation of an object by a device separated from it by some distance utilizing the characteristics response of different objects to emissions in the electromagnetic energy is measured in a number of spectral bands for the purpose of identification of the object.

In such study single tabular form of data or map data is not sufficient enough which can provide can be, combined with information's obtained from existing maps and tabular data.

- Remote Sensing techniques using various plate form has provide its utility in agricultural survey

- Satellite data provides the actual synoptic view of large are at a time, which is not possible from conventional survey methods.
- The process of data acquisition and analysis is very fast through Geographic Information System (GIS) as compared to conventional methods.

Remote Sensing techniques have a unique capability of recording data in visible as well as invisible (i.e. ultraviolet, reflected infrared, thermal infrared and microwave etc.) part of electromagnetic spectrum. Therefore certain phenomenon, which cannot be seen by human eye, can be observed through remote sensing techniques i.e. the trees, which are affected by disease, or insect attack can be detected by remote sensing techniques much before human eyes see them.

GENERATING AGRICULTURAL DATA AND ITS PROBLEMS

The present system of agricultural data is collected throughout the nation. The main responsibility of collection agricultural survey lies on the Director of Land Records, Director of agriculture and District Statistical Office under the Ministry of Agriculture.

These data are collected not only on a local but also some extent of district and state level. The associate of agricultural survey on crops (crop production, type of crop and crop yield), range land (condition of range, forest type, water quality, types of irrigation system and soil characteristics) and livestock (livestock population, sex of animal, types of farm and distribution of animals).

The basic problems in this survey are;

- Reliability of data
- Cost and benefits
- Timeless
- Incomplete sample frame and sample size
- Methods of selection
- Measurement of area
- Non sampling errors
- Gap in geographical coverage
- Non availability of statistics at disaggregated level.

Remote Sensing techniques make it use before the remote sensing data may provide solution to these particular problems of agricultural survey.

Remote Sensing Techniques for Agricultural Survey

The given factors influenced the use of remote sensing in agricultural surveys; via 1. Characteristics of the agricultural landscape 2. Characteristic of EMR on Agricultural survey. Detection, identification, measurement and monitoring of agricultural phenomena are predicated on the assumption that agricultural landscape features (e.g. crops, livestock, crop infestations and soil anomalies) have consistently identifiable signatures on the type of remote sensing data.

Some of the parameters which may cause these identifiable signatures include crop type, state of maturity, crop density, crop geometry, crop vigor, crop moisture, crop temperature, soil moisture, soil temperature. An image analysis can correlate a certain signature with one of these many characteristics. Remote Sensing techniques in agriculture survey which affect the signature on remote sensing imagery. It is important to consider briefly the significance of choosing the appropriate sensor system, as well as the scale and resolution requirements that will yield optimum benefits for objectives of agricultural survey.

SIGNATURE IN REMOTE SENSING

The knowledge of spectral signatures is essential for exploiting the potential of remote sensing techniques. This knowledge enables one to identify and classify the objects of agricultural resources. It is also required for interpretation of all remotely sensed data, especially in agricultural resource data whether the interpretation is carried out visually or using digital techniques. It also helps us in specifying requirements for any remote sensing mission e.g. which optimal wave length bands to be used or which type of sensor will be best suited for a particular task (agricultural survey). All objects of agricultural resource on the surface of the earth have characteristic spectral signatures. The average spectral reflectance curves (or) spectral signatures for three typical earth's features; vegetation, soil and water.

The spectral reflectance curves for vigorous vegetation manifests the "Peak-and valley" configuration. The valleys in the visible portion of the spectrum are indicative of pigments in plant leaves. Dips in reflectance that can be sent at wavelengths of 0.65 mm, 1.4 mm and 1.9 mm are attributable to absorption of water by leaves. The soil curves show a more regular variation of reflectance. Factors that evidently affect soil reflectance are moisture content, soil texture, surface roughness and presence of organic matter. The water curves shows that from about 0.5 mm, reduction in reflectance with increasing wavelength, so that in the near infrared range, the reflectance of deep clear water is virtually zero (Mather, 1987) However, the spectral reflectance of water is significantly affected by the presence of dissolved and suspended organic and inorganic material and by the depth of the water body. Determinations of spectral signatures implies basic under standing of interaction of electromagnetic radiation with agricultural resources objects. This is also necessary for analysing and designing sensor systems for agricultural survey.

Sensor systems in Remote Sensing

In remote sensing the acquisition of data is depending upon the sensor system used. Various remote sensing platforms (Aircraft, Satellite) are equipped with different sensor systems. Sensor is a device that receives electromagnetic radiation, converts it into a signal and presents it in a form suitable of obtaining information about the land or earth resource as used by an information gathering system. Sensor can be grouped, either on the basic of energy source. They are as classified.

Active sensor: An active sensor operates by emitting its own energy, which is needed to detect the various phenomena (e.g. RADAR, camera with a flash gun).

Passive sensor: The operation of passive sensor is dependent of the existing sources of energy, like sun (e.g. photographic systems, multispectral scanners).

The given sensor system of camera are in agricultural survey.

Photographic Cameras

The photographic system, having conventional camera with black and white photography, is the oldest and probably, so far, the most widely used sensor for recording information about ground object.

Photographic cameras have been successfully used in aircraft platform remote sensing. In this system, the information is limited to size and shape, as the films used are sensitive only to visible region of spectrum. The response of black & white films is about 0.4-0.7 mm for infrared imagery, films with response extending up to 0.9 mm are available.

Return Beam Vidicon (RBM)

This is very similar to a television camera. In such a system, a fixed camera lens on a photosensitive semi-transparent sheet forms the ground image.

This image is created on the surface as electrical change or potential. The TV cameras are the best example of high resolution, operated in space for resource survey was the RBV used in LAND SAT series. On LAND SAT I, II and III RBV cameras were used, each corresponding to a different wavelength band 0.475-0.585 mm (green), 0.580-0.690 mm (red) and 0.690-0.830 mm (near infrared).

The Indian experimental remote sensing satellite, Bhaskara-I and II carried a two-band TV camera system, Multispectral imagery was produced in LAND SAT and Bhaskara by using separate camera tubes of each band and selecting the spectral band with appropriate filters.

Optical-mechanical Scanners

This imaging system has the advantage that any set of desired spectral bands can be selected with appropriate filter and detector combinations. The mostly widely used sensor in this category is the MSS on LAND SAT series. MSS has four spectral bands, covering form 0.5-to 1.1 mm region. MSS operates on the principle of scanning successive lines at right angles to the flight path by means of a rotation or oscillating optical system. The radiation levels along the lines are recorded by appropriate sensor elements. When used in the visible band, the collected light can be split by the optics and separately filtered and recorded, giving simultaneous multispectral recording from the one instrument. MSS can record in any part of ultraviolet to near IR window. They are use also in the thermal IR windows.

Radar and Microwave Sensors

The acquisition of data in microwave region has been possible since 1950s but its application to natural resources is considerably less developed, as compared to the visible and IR image interpretations. Microwave sensors have distinct advantages because they are unaffected by atmospheric conditions and are thus able to penetrated smoke, clouds, haze and snow. Under this system, Plan Position Indicator (PPI), Side Looking Air borne Radar (SLAR) and Synthetic Aperture Radar (SAR) can be grouped. These systems offer day and night as well as all weather capability and ability to penetrate a cover of vegetation.

Advance Remote Sensors

Linear Imaging and Self Scanning Sensors (LISS) are the advanced imaging systems. This type of scanning sensor are used an array of solid-state devices. The array may be made of photo-diodes, phototransistors or Charge-Coupled

Devices (CCDs). In the LISS, the optics focuses a strip of terrain in the cross-track into the sensor array.

The image from each detector is stored and shifted out sequentially to receive a video signal. The SPOT (Satellite Probatorie d' Observation de la Terra) and IRS (Indian Remote Sensing Satellite) series carry such solid-state sensor systems, which are also known as push-broom scanners.

The IRS IC most advanced satellite, carries an improved sensor system. Besides carrying a sophisticated LISS-III camera, it has a Panchromatic camera (PAN) and a Wide Field Sensor (WiFS).

The PAN has been designed to provide data with a spatial resolution of 5.8m in stereo mode, with a ground swath of 70km, whereas WiFS provides data in two spectral bands, with a spectral resolution of 188m and a ground swath of 180km.

PRACTICAL PROBLEMS IN INDIAN AGRICULTURE

Precision Agriculture has been mostly confined to developed countries. Reasons of limitations of its implementation in developing countries like India are:

a. Small land holdings,
b. Heterogeneity of cropping systems and market imperfections,
c. Lack of technical expertise knowledge and technology (India spends only 0.3% of its agricultural Gross Domestic Product in Research and Development)
d. High cost.

In India, major problem is the small field size. More than 58 percent of operational holdings in the country have size less than 1ha. Only in the states of Punjab, Rajasthan, Haryana and Gujarat more than 20 per cent of agricultural lands have operational holding size of more than four hectare. There is a scope of implementing precision Agriculture for crops like, rice and wheat especially in the states of Punjab and Haryana. Commercial as well as Horticultural crops shows a wider scope for precision Agriculture. In India, broadly two types of agriculture viz., high input agriculture characterized by the provision of assured irrigation and other agricultural inputs, and subsistence farming, which is confined mostly to rain-fed, or dry land regions, are prevalent. Nearly two-third arable land in India is rain-fed. The crop yields are very low (˜ 1 t ha-1) and very good potential exists for increasing productivity of rain-fed Cropping systems.

Methodology

Using a GPS along with a digital drainage map, the farmer is able to apply these pesticides in a safer manner. The spraying equipment can be preprogrammed automatically to turn off when it reaches the distance limitation or zone of the drainage feature. Additionally, farmers can preprogram the rate of pesticide or fertilizer to be applied so that only the amount needed determined by the soil condition is applied varying this rate from one area of the field to another. This saves money and allows for safer use of these materials.

Data collected by the GPS operations can be automatically recorded with the GIS program. Remotely sensed data can be analysed and added to the GIS using soil maps, digital terrain and field operations information as ground truth. Additionally the farmers are able to record observations through the growing season such as weed growth, unusual plant stress or coloring and growth conditions. This can be used to guide further field operations like spraying, fertilizing and irrigating and it is part of the permanent record. Precision farming will make a strong impact on the way farmers manage their farm operations in the future.

Seed Spacing

The yield from the seeds is maximum when adequate care is taken to adequately space them to enable each plant the optimum requisites for its growth, namely the soil nutrients, water, sunlight and protection against pest infestation. This is best accomplished by varying the seeding rate even within a single piece of land according to the soil conditions such as texture, organic matter and available soil moisture. One would plant fewer seeds in sandy soil as compared to silt loam soils because of less available moisture.

The lower seed population usually has larger heads (ears) of harvested seeds providing for a maximum yield. Since soils vary even across an individual farm field, the ability to change seeding rates as one goes across the field allows the farmer to maximize this seeding rate according to the soil conditions. A computerized soil map of a specific field on a computer fitted on the tractor along with a GPS can tell farmers where they are in the field allowing the opportunity to adjust this seeding rate as they go across their fields.

STEPS TO BE TAKEN FOR IMPLEMENTING PF

In the present existent situation, the potential of precision agriculture in India is limited by the lack of appropriate measurement and analysis techniques for agronomically important factors (National Research Council, 1997). High accuracy sensing and data management tools must be developed and validated to support both research and production. The limitation in data quality/ availability has become a major obstacle in the demonstration and adoption of the precision technologies. The adoption of precision agriculture needs combined efforts on behalf of scientists, farmers and the government. The following methodology could be adopted in order to operationalise precision farming in the country.

1. Creation of multidisciplinary teams involving agricultural scientists in various fields, engineers, manufacturers and economists to study the overall scope of precision agriculture.
2. Formation of farmer's co-operatives since many of the precision agriculture tools are costly (GIS, GPS, RS, etc.).
3. Government legislation restraining farmers using indiscriminate farm inputs and thereby causing ecological/environmental imbalance would induce the farmer to go for alternative approach.
4. Pilot study should be conducted on farmer's field to show the results of precision agriculture implementation.
5. Creating awareness amongst farmers about consequences of applying imbalanced doses of farm inputs like irrigation, fertilizers, insecticides and pesticides.

Realizing the potential of space technology in precision farming, the Department of Space, Government of India has initiated eight pilot studies in well-managed agricultural farms of the ICRISAT, the Indian Council of Agricultural Research and the Agricultural Universities, as well as in farmers' fields. The pilot studies aim at delineating homogeneous zones with respect to soil fertility and crop yield, estimation of potential yield, yield gap analysis, monitoring seasonally-variable soil and crop conditions using optical and microwave sensor data, and matching the farm inputs to bridge the gap between potential and actual yield through Spatial Decision Support Systems (SDSS). The test sites are spread over a fairly large area across a cross section of agro-climatic zones of the Indian sub-continent, and cover some of the important crops like wheat, rice, sorghum, pigeon pea, chickpea, soybean and groundnut.

The next step would be to generate detailed-level information on soil resources addressing potentials and limitations of individual fields since except for states like Punjab, Haryana, Madhya Pradesh and Maharashtra where fields size is quite large, practically individual field could be treated as a homogenous management unit for the purpose of precision farming.

DEFINITION OF PRECISION AGRICULTURE

"Precision Farming is the title given to a method of crop management by which areas of land/crop within a field may be managed with different levels of input depending upon the yield potential of the crop in that particular area of land. The benefits of so doing are two fold:

- The cost of producing the crop in that area can be reduced and,
- The risk of environmental pollution from agrochemicals applied at levels greater than those required by the crop can be reduced".

Precision farming is an integrated agricultural management system incorporating several technologies. The technological tools often include the *global positioning system, geographical information system, yield monitor, variable rate technology, and remote sensing*. The *global positioning system* ("GPS") is a network of satellites developed for and managed by the U.S. Defence Department. The GPS constellation of 24 satellites orbiting the earth, transmit precise satellite time and location information to ground receivers. The ground receiving units are able to receive this location information from several satellites at a time for use in calculating a triangulation fix thus determining the exact location of the receiver. A *geographical information system* ("GIS") consists of a computer software data base system used to input, store, retrieve, analyze, and display, in map like form, spatially referenced geographical information.

Goal of Precision Farming

The goal of precision farming is to gather and analyze information about the variability of soil and crop conditions in order to maximize the efficiency of crop inputs within small areas of the farm field. To meet this efficiency goal the variability within the field must be controllable.

Efficiency in the use of crop inputs means that fewer crop inputs such as fertilizer and chemicals will be used and placed where needed.

The benefits from this efficiency will be both economical and environmental. Environmental costs are difficult to quantify in monetary terms. The reduction of soil and groundwater pollution from farming activities has a desirable benefit to the farmer and to society.

Research

Precision farming is an integration of several technologies. U.S., and foreign governments originally paid for the development and support of technologies such as GPS, Remote Sensing, and GIS, for military or other civilian purposes long before the emergence of precision farming. It is the use of these advanced technologies that has generated enormous amounts of data to process with computers. A very basic question that still needs to be answered by researchers is what does all of this data mean in order for farmers to make profitable management decisions? In other words we have the technology to gather the data from the field but we don't yet have the knowledge to transform the data into answers for agricultural management decisions.

Universities around the world, and foreign and U.S., government entities such as the Agricultural Research Stations of the USDA are conducting extensive research on precision farming. Research projects will apply technologies such as remote sensing, GPS, GIS, and VRT to create management decision support systems. A goal of many publicly funded research institutions is to promote technology transfer from government agencies to the private sector.

CONTENT ISSUES AND THE GATEWAY DEBATE

Despite all the natural advantages, India's productivity of food grains per hectare is no more than three-fourths of the world average and less than half of that in agriculturally advanced countries, per capita food grain availability even after the Green Revolution, has been less than two thirds of the world average. Only five states in India, namely Himachal Pradesh, Punjab, Haryana, Uttar Pradesh and Madhya Pradesh – produce more grain than their populations can consume. The combined population of the five states is less than one-third of the total of the country. More than two thirds of the population lives in states that are still food-deficit. This requires transport of lakhs of tonnes of food grain, involving high costs and pilferage. Our effort should have been to make all the states self-sufficient with respect to food grains and if some disturbances occurred due to unnoticed natural calamities the nation must be in an ever ready position to mitigate such challenging tasks.

The Indian green revolution is also associated with negative ecological/ environmental consequences. The status of Indian environment shows that, in India, about 182 million ha of the country's total geographical area of 328.7 million ha is affected by land degradation of this 141.33 million ha are due to water erosion, 11.50 million ha due to wind erosion and 12.63 and 13.24 million ha due to water logging and chemical deterioration (salinization and loss of nutrients) respectively. On the other end, India shares 17 per cent of world's population with 2.5 per cent of geographical area, 1 per cent of gross world product, 4 per cent of world carbon emission and hardly 2 per cent of world forest area. The Indian status on environment is though not alarming when compared to developed countries, it gives an early warning to take appropriate precautionary measures. The growth rate of grain production during the Ninth plan has been less than the population growth rate. The poor agricultural performance has not been because of the vagaries of monsoon. From 1997-98 to 2002-03, rainfall was between 92 and 106 per cent of the normal. Per capita availability of grain and per capita calorie intake, which were less than the minimum required for adequate nutrition, have further declined. According to Human Development Report 2003, the percentage of the undernourished in India, which was 21 a few years ago, has now reached 24. The Government claims that India has emerged as the seventh largest exporter of food grains in the world. This is nothing to be proud of, if we take into account that the total Indian grain export in 2002-03 did not add up to even 4 per cent of the total world exports; and the value of our grain exports did not add up to even the value of our imports of vegetable oils and pulses. The more crucial question, however, is whether it is morally justifiable to export grain when 24 per cent of the population remains under-nourished. The decline in agricultural growth and increase in rural poverty have been due to the long persisting government indifference towards the farm sector, which is evident from plan outlays on Agricultural and Allied activities, Rural Development and Irrigation, which added up to 37.1 per cent of the total during the first plan were brought down to only 19.4 per cent during the Ninth plan. However, the main reason, for poor performance of the farm sector has been the long persisting adverse terms of trade policies for agriculturists in addition to the mismanagement of natural resources leads to ever ending crisis.

THE NEED FOR PRECISION FARMING

The 'Green revolution' of 1960's has made our country self sufficient in food production. In 1947, the country produced a little over six million tonnes of wheat, in 1999; our farmers harvested over 72 million tonnes, taking the country to the second position in wheat production in the world. The production of food grains in five decades, has increased more than three fold, the yield during this period has increased more than two folds. All this has been possible due to high input application, like increase in fertilization, irrigation, pesticides, higher use of HYV's, increase in cropping intensity and increase in mechanization of agriculture.

Fatigue of Green Revolution

Green revolution of course contributed a lot. However, even with the spectacular growth in the agriculture, the productivity levels of many major crops are for below than expectation. We have not achieved even the lowest level of potential productivity of Indian high yielding varieties, whereas the worlds highest productive country have crop yield levels significantly higher than the upper limit of the potential of Indian HYV's. Even the crop yields of India's agriculturally rich state like Punjab is far below than the average yield of many high productive countries.

Natural Resource Degradation

The green revolution is also associated with negative ecological/environmental consequences. The status of Indian environment shows that, in India, about 182 million ha of the country's total geographical area of 328.7 million ha is affected by land degradation of this 141.33 million ha are due to water erosion, 11.50 million ha due to wind erosion and 12.63 and 13.24 million ha are due to water logging and chemical deterioration (salinisation and loss of nutrients) respectively. On the other end, India shares 17 per cent of world's population, 1 per cent of gross world product, 4 per cent of world carbon emission, 3.6 per cent of CO_2 emission intensity and 2 per cent of world forest area. One of the major reasons for this status of environment is the population growth of 2.2 per cent in 1970 – 2000. The Indian status on environment is, though not alarming when compared to developed countries, gives an early warning. In this context, there is a need to convert this green revolution into an evergreen revolution, which will be triggered by farming systems approach that can help to produce more from the available land, water and labour resources, without either ecological or social harm. Since precision farming, proposes to prescribe tailor made management practices, it can help to serve this purpose.

OPERATIONALIZATION OF PRECISION FARMING

"Agriculture is the backbone of the Indian Economy"-said Mahatma Gandhi five decades ago. Even today, as we enter the new millennium, the situation is still the same, with almost the entire economy being sustained by agriculture, which is the mainstay of the villages. Not only the economy, but also every one of us looks up to agriculture for our sustenance too. Therefore, it is no surprise if agriculture gets the celebrity status in the name of Precision farming (PF).

In recent times, the researchers in the field have been busy formulating methodologies and fabricating new implements for precision farming. It is here the challenge arises considering the implementation of the technology at various levels in the Global community. The need of the hour is not application of the technology but the adoption of appropriate technology, which would suit the particular level of the global community. In India, the farming practices are too haphazard and non-scientific and hence need some forethought before implementing any new technology. It is here that this chapter comes out with a cost-benefit analysis to prove that precision farming is possible in India. Applications of agricultural inputs at uniform rates across the field without due regard to in-field variations in soil fertility and crop conditions does not yield desirable results in terms of crop yield. The management of in-field variability in soil fertility and crop conditions for improving the crop production and minimizing the environmental impact is the crux of precision farming. Thus, the information on spatial variability in soil fertility status and crop conditions is a pre-requisite for adoption of precision farming.

Space technology including global positioning system (GPS) and GIS holds good promise in deriving information on soil attributes and crop yield, and allows monitoring seasonally-variable soil and crop characteristics, namely soil moisture, crop phenology, growth, evapotranspiration, nutrient deficiency, crop disease, and weed and insect infestation, which, in turn, help in optimizing inputs and maximizing crop yield and income. Though widely adopted in developed countries, the adoption of precision farming in India is yet to take a firm ground primarily due to its unique pattern of land holdings, poor infrastructure, lack of farmers' inclination to take risk, socio-economic and demographic conditions. The aim of this chapter is to suggest measures for the implementation of this novel technique in the country with a greater emphasis on the systematic approach towards its operationalisation.

Precision Farming

The conventional agronomic practices follow a standard management option for a large area irrespective of the variability occurring within and among the field. For decades now, the farmers have been applying fertilizers based on recommendations emanating from research and field trials under specific agro-climatic conditions. Since soil-nutrient, characteristics vary not only from one region to another, but also from field to field. Even within a field, there is a need to take into account such variability while applying fertilizers to a particular crop. Consideration of in-field variations in soil fertility and crop conditions and matching the agricultural inputs like seed, fertilizer, irrigation, insecticide, pesticide, etc. in order to optimize the input or maximizing the crop yield from a given quantum of input, is referred to as precision farming or precision agriculture or precision crop management. The term "precision farming" means carefully tailoring the soil and crop management to fit the different conditions found in each field. It is defined as the application of technologies and principles to manage spatial and temporal variability associated with all aspects of agricultural production.. It is also referred to as "prescription farming", "site specific farming" or "variable rate technology."

By catering to this variability, called precision farming, one can improve the productivity or reduce the cost of production and diminish the chance of environmental degradation caused by excess use of inputs. Thus, mapping and analysis of within field variability is an essential input for precision crop management.

Thus, PF involves acquiring the variations in crop or soil properties, mapping, and analysing the variations, adopting suitable management techniques to maximize the yield. Farmers have been applying fertilizers based on recommendations emanating from research and field trials under specific agro-climatic conditions, which have been extrapolated to a regional level. Since soil nutrient characteristics vary not only between regions and between farms but also from plot to plot, and within a field or plot, there is a need to take into account such variability while applying fertilizers to a particular crop. Consideration of in-field/plot variations in soil fertility and crop conditions and matching the agricultural inputs like seed, fertilizer, irrigation, insecticide, pesticide, etc. in order to optimize the input or maximizing the crop yield from a given quantum of input, is referred to as precision farming or precision agriculture or precision crop management.

The information for variability map can be obtained from soil tests for nutrient availability, yield monitors for crop yield, soil samples for organic matter content, information in soil maps, or ground conductivity meters for soil moisture. Generally, the fields are manually sampled along a regular grid and the analysed results of the samples are interpolated using geostatistical techniques. These techniques are time consuming, labour intensive and in many cases destructive especially, for agricultural situation in India. With small size of landholdings and low income of farmers, the adoption of this methodology in its present form is not feasible. Various workers have shown the advantages of using remote sensing technology to obtain spatially and temporally variable information for precision farming. In an earlier work, Ray *et al.* have shown the usefulness of IRS merged data in mapping the variability.

7

Precision Farming and Agriculture

Ever since the man appeared on the earth, he has been harnessing the natural resources to meet his basic requirements. Reference to soil, water and air as basic resources, their management and means to keep them pure are mentioned in the Vedas, Upanishads and in ancient Hindu literature. The phenomenal increase in population of both man and animal in the last century and fast growing industrialization and urbanization in last few decades have overstrained the natural resource base, which are getting degraded much faster than ever before. Thus, the attention of whole world is focused on how to increase production to feed the burgeoning population and the question uppermost in every ones mind is "Can we produce enough food in a sustainable manner without damage to the natural resource base?"

Belying all predictions made to the contrary, India could achieve unprecedented increase in the food grain production as a result of expansion of irrigation and technological advancement in agriculture. While it has been a satisfying experience, Indian agriculture would need a new vision to make rapid progress in the ensuing millennium. To achieve the required growth will not be easy as some of the existing production systems are based on unsustainable use of the resources. The signs of fatigue in the natural resources have already appeared which is a cause for serious concern to the planners, decision-makers and researchers alike.

Precision Agriculture is the application of technologies and principles to manage spatial and temporal variability associated with all aspects of agricultural production for improving production and environmental quality. The success in precision agriculture depends on the accurate assessment of the variability, its management and evaluation in space-time continuum in crop production. The agronomic feasibility of precision agriculture has been intuitive, depending largely on the application of traditional arrangement recommendations at finer scales. The agronomic success of precision agriculture has been quite convincing in crops like sugar beet, sugarcane, tea and coffee. The potential for economic, environmental and social benefits of precision agriculture is largely unrealized because the space-time continuum of crop production has not been adequately addressed.

Successful implementation of precision agriculture depends on numerous factors, including the extent to which conditions within a field are known and manage, the adequacy of input recommendation and the degree of application control. The enabling technologies of precision agriculture can be grouped in to fine major categories: Computers, Global Positioning System (GPS), Geographic Information System (GIS), Remote Sensing (RS) and Application control. Aspects of precision agriculture encompass a broad array of topics including variability of the soil resource base, weather, plant genetics, crop diversity, machinery performance and most physical, chemical and biological inputs used in crop production. Precision agriculture must fit the needs and capabilities of the farmer and must be profitable.

THE BASIC COMPONENTS OF PRECISION FARMING

Precision farming basically depends on measurement and understanding of variability, the main components of precision farming system must address the variability. Precision farming technology enabled, information based and decision focused, the components include, (the enabling technologies) Remote Sensing (RS), Geographical Information System (GIS), Global Positioning System (GPS), Soil Testing, Yield Monitors and Variable Rate Technology. Precision farming requires the requisition, management, analysis and output of large amount of spatial and temporal data. Mobile computing systems were needed to function on the go in farming operations because desktop systems in the farm office were not sufficient. Because precision farming is concerned with spatial and temporal variability and it is information based and decision focused. It is the spatial analysis capabilities of GIS that enable precision agriculture. GPS, DGPS has greatly enabled precision farming and of great importance to precision farming, particularly for guidance and digital evaluation modelling position accuracies at the centimeter level are possible in DGPS receivers. Accurate guidance and navigation systems will allow for farming operations at height and under unfavorable weather conditions even.

In India, we have all these technologies available and they can be implemented through agricultural training centers by giving training to agriculture officers in these technologies.

BASIC STEPS IN PRECISION FARMING

The basic steps in precision farming are,

i. Assessing variation

ii. Managing variation and

iii. Evaluation.

The available technologies enable us in understanding the variability and by giving site specific agronomic recommendations we can manage the variability that make precision agriculture viable. And finally evaluation must be an integral part of any precision farming system. The detailed steps involved in each process are clearly depicted in a diagramme.

Assessing Variability

Assessing variability is the critical first step in precision farming. Since it is clear that one cannot manage what one does not know. Factors and the processes that regulate or control the crop performance in terms of yield vary in space and time. Quantifying the variability of these factors and processes and determining when and where different combinations are responsible for the spatial and temporal variation in crop yield is the challenge for precision agriculture.

Techniques for assessing spatial variability are readily available and have been applied extensively in precision agriculture. The major part of precision agriculture lies in assessing to spatial variability. Techniques for assessing temporal variability also exist but the simultaneous reporting a spatial and temporal variation is rare. We need both the spatial and temporal statistics. We can observe the variability in yield of a crop in space but we cannot predict the reasons for the variability. It needs the observations at crop growth and development over the growing season, which is nothing but the temporal variation. Hence, we need both the space and time statistics to apply the precision farming techniques. But this is not common to all the variability/ factor that dictate crop yield. Some variables are more produced in space rather with time, making them more conducive to current forms of precision management.

Managing Variability

Once variation is adequately assessed, farmers must match agronomic inputs to known conditions employing management recommendations. Those are site specific and use accurate applications control equipment. We can use the technology most effectively.

In site-specific variability management. We can use GPS instrument, so that the site specificity is pronounced and management will be easy and economical. While taking the soil/plant samples, we have to note the sample site coordinates and further we can use the same for management. This results in effective use of inputs and avoids any wastage and this is what we are looking for.

The potential for improved precision in soil fertility management combined with increased precision in application control make precise soil fertility management as attractive, but largely unproven alternative to uniform field management. For successful implementation, the concept of precision soil fertility management requires that within-field variability exists and is accurately identified and reliably interpreted, that variability influences crop yield, crop quality and for the environment. There fore inputs can be applied accurately.

The higher the spatial dependence of a manageable soil property, the higher the potential for precision management and the greater its potential value.

The degree of difficulty, however, increases as the temporal component of spatial variability increases. Applying this hypothesis to soil fertility would support that Phosphorus and Potassium fertility are very conducive to precision management because temporal variability is low. For N, the temporal component of variability can be larger than its spatial component, making precision N management much more difficult in some cases.

Evaluation

There are three important issues regarding precision agriculture evaluation.

- Economics
- Environment and
- Technology transfer.

The most important fact regarding the analysis of profitability of precision agriculture is that the value comes from the application of the data and not from the use of the technology. Potential improvements in environmental quality are often cited as a reason for using precision agriculture. Reduced agrochemical use, higher nutrient use efficiencies, increased efficiency of managed inputs and increased production of soils from degradation are frequently cited as potential benefits to the environment. Enabling technologies can make precision agriculture feasible, agronomic principles and decision rules can make it applicable and enhanced production efficiency or other forms of value can make it profitable.

The term technology transfer could imply that precision agriculture occurs when individuals or firms simply acquire and use the enabling technologies. While precision agriculture does involve the application of enabling technologies and agronomic principles to manage spatial and temporal variability, the key term is manage. Much of the attention in what is called technology transfer has focused on how to communicate with the farmer. These issues associated with the managerial capability of the operator, the spatial distribution of infrastructure and the compatibility of technology to individual farms will change radically as precision agriculture continues to develop.

Technology Transition

Precision agriculture is dependent on the existence of variability in either or both product quantity and quality. If this variability does not exist then a uniform management system is both the cheapest and most effective management strategy and precision farming is redundant. Thus, in precision farming, "Variability of production and quality equals opportunity". Having said this, the nature of the variation is also important in determining the potential for PA in a system. For example the magnitude of the variability may be too small to be economically feasible to manage. Alternatively the variability may be highly randomized across the production system making it impossible to manage with current technology. Finally the variability may due to a constraint that is not manageable. Thus the implementation of precision farming is limited by the ability of current variable rate technology (VRT-machinery/ technology that allows for differential management of a production system) to cope with the highly variable sites and the economic inability to produce returns from sites with low variability using precision farming (VRT).

GEOGRAPHICAL INFORMATION SYSTEM

The Geographic Information System (GIS) contributes significantly to precision farming by allowing presentation of spatial data in the form of a map. In addition, GIS forms an ideal platform for the storage and management of model input data and the presentation of model results, which the process model provides.

DGPS

'Do the right thing, in the right place and in right time'

This is where GPS comes into picture. In addition, the accuracy, which is the important factor in PF, demands for DGPS (Differential Global Positioning System). GPS makes use of a series of military satellites that identify the location of farm equipment within a meter of an actual site in the field. The value of knowing a precise location within inches is that:

1. Locations of soil samples and the laboratory results can be compared to a soil map,
2. Fertilizer and pesticides can be prescribed to fit soil properties (clay and organic matter content) and soil conditions (relief and drainage),
3. Tillage adjustments can be made as one finds various conditions across the field, and
4. One can monitor and record yield data as one goes across the field.

The Global Positioning System (GPS) technology provides accurate positioning system necessary for field implementation of variable rate technology (VRT). The Internet makes possible the development of a mechanism for effective farm management using remote sensing.

Variable Rate Applicator

The variable rate applicator has three components:

1. Control computer
2. Locator and
3. Actuator.

The control computer coordinates the field operation. It has a map of desired activity as a function of geographic location. It receives the equipment's current location from the locator, which has a GPS in it, and decides what to do based upon the map in its memory or data storage.

It then issues the command to the actuator, which does the input application.

THE BUSINESS SIDE OF FARMING

Farming as a Business

Farming has not usually been considered a business. The diversity of the duties of the farmer, the area over which the operations of the farm extend and the complexity of the records required, combine in making difficult the organization of the details of farming into business form. The successful financial operation of a farm presents quite as complex problems and calls for at least as much business ability and judgment as is required in operating a store with the same investment. Farming, therefore, should be considered as a business, and the man who can produce his crops and products at the lowest cost and sell them at the highest price, investing the proceeds to the best advantage, should be considered the best farm manager. The man who knows the details of the cost of production and operation, and whose records show the profitable and unprofitable lines of production, thus enabling him to eliminate those that do not yield a profit, may be counted as the best business man.

A farmer should know the elements of soil fertility. He must understand the principles of the movement of soil water, and the action of soil bacteria. He should understand the nature of plant growth and be familiear with varieties and species of plants and with the effect of one crop on the crop following. He must also be familiar with animals and their habits and know how to feed and care for them. In addition, he must know how to buy and sell to advantage, make contracts, and plan his buildings and his farm so as to necessitate a light expenditure for labour, also that he may distribute his labour to advantage over the various farm enterprises. And he should know how to keep accounts. The farmer in organizing his business could well follow the example of the merchant. The merchant first takes an inventory of his stock. He studies the demand for his goods, both present and prospective. He notes the supply, the cost, and the demand for each article.

He calculates the labour required to operate his business and such other items of expense are considered as may be legitimately charged against the business. He regulates his purchases and his prices according to the cost of securing his goods and putting them on the market. In conducting a large store business, it is customary to organize it into departments, putting some competent person in charge of each department and having the labour and accounting charges so systematized and recorded as to show the profit or loss from each department and from the business as a whole.

The farmer should likewise take an inventory of his capital, stock, and equipment. He should consider the type of farming to which the soil and climate are adapted. He should consider the fertility of the soil and the demand that will be made upon it by the crops grown.

He should consider, in connection with the soil fertility, the sources from which it may be renewed and at what cost. He must study the markets, the transportation, and the demand for such crops as he grows; also the cost of producing each of the crops and the probable net profit that will be returned. His labour likewise should be charged against the various crops or enterprises and distributed to the best advantage.

In studying the problems of farm organizations, interest on investment, taxes, insurance, and other expense must be included as they affect the financial result. As in a large store business, it is frequently necessary to organize the large farm into departments, keeping accounts with the dairy, with the swine, the grain crops, the garden, and other similar enterprises. Where the business is large enough, it is well to put an expert in charge of each large branch or group of enterprises, thus enabling one to use cheaper labour for performing the work or making the labour more effective by closer supervision.

Where the farming is conducted as an organized business, and accounts are kept with the various lines of work, it is possible at the end of the year, to make a business statement which will show which lines have been profitable. The manager then can change his methods or drop out those lines that prove to be unprofitable and the business as a whole may be put on a better basis.

Investment

The investment of money in land, buildings, and equipment demands careful consideration. It is possible to pay so much for a farm that it will be impossible to produce sufficient revenue to meet the expense of operation and to pay a normal rate of interest on the money invested. This is particularly true where low-priced products are produced. A farm may be highly productive but so located that it will be impossible to market the produce on a profit bearing basis. One should study closely the market facilities of the neighbourhood and raise supplies which can be successfully marketed locally, or which can be transported to a market that pays well for such produce. Unless the produce of the farm is well related to the market, the farm is likely to be operated at a loss.

Proportion in Real Estate

It is a mistake for one to invest all of his capital in the real estate itself. Sufficient capital should be reserved for operating the farm. The hunger for land has induced many farmers to buy more land that they can equip and operate well. Such farmers are said to be land poor. They would secure greater profit from a medium sized farm, well tilled and managed, than from a large one which is insufficiently equipped and poorly operated. Rarely should more than 50 or 60 per cent of the capital be tied up in the land. The size of the farm and the amount of equipment are closely related to the possible profits. A farm of forty or eighty acres devoted to diversified crops and live stock, cannot be so economically equipped per acre as a larger farm. The investment per acre in machinery will be higher as the cost will be spread over fewer acres

the total annual cost would be only $220, charging the same interest and expense rates as in the first instance, and assuming that the rate of depreciation would be the same. While the $4000 barn would undoubtedly be a better barn, it would not add to the production of the cows housed, unless it was much more comfortable. It would not add to the net profit from the investment unless the labour of doing the chores and caring for the cows would be considerably reduced by greater convenience. The cost of horse barns, swine barns, and other buildings can be similarly calculated. One should not erect a building unless it is going to add to the efficiency of the live stock, shelter hay or machinery, or lessen the labour of doing the chores. It is wiser to invest money in drainage or in better tillage of the soil, than to invest it in buildings that shelter unproductive stock, or that add nothing to the earning power of the farm.

Cost of Labour

The employment, organization and direction of labour demands considerable study. The value of a good farm manager lies quite as much in his abilitiy so to select and direct labour as to yield a profit, as it does in his ability to drive a good bargain or sell his crops well. The only reason for employing labour is to increase the product and consequent profit. If a farmer can, by employing a man eight months in the year at $40 per month, increase the product of his farm by $500, he will be warranted in employing the labour. If, however, the $320 invested in labour should yield an increase of only $200 in the products of the farm, employment would be at a loss.

The Factors of Production

Three primary factors are necessary in agricultural production. These are capital, land, and labour. The adjustment of these three factors is an important part of the business of the farm owner or manager, and determines largely the profits that may be made from the individual farm.

Capital, as commonly understood, includes the money value represented in the investment of the farm property, no matter what the form may be. Implements, live stock, teams, buildings, and other articles of equipment, are each a part of the capital of the farm. Cash for operating is also included.

Land represents the larger part of capital on most farms, and demands special consideration because the amount of land available for agricultural purposes is limited. Farmers have for this reason, regarded it wise to secure large quantities in localities where it was cheap, anticipating a rise in value. Location and demand for land in particular sections has led to much speculation, and land values fluctuate frequently.

The proportionate investment in each of the three forms,—circulating capital, land, and labour—bears a vital relation to the profits possible from the farm, and must be given the most careful consideration by the person who is buying and equipping a farm.

Capital Classified

There are two forms of capital in common use. They are known as fixed or invested capital, and circulating or working capital. The fixed capital properly includes all forms of permanent equipment, such as investment in land, buildings, implements, teams, and other articles that are used continuously. In land it includes the natural value and the value of the improvements that have been made upon it.

Picking stones from a rough section of land adds to its value and increases the capital invested. Clearing trees from the land has the same effect. Wells, drainage, roads, fences, and other forms of improvement which are permanent and which become part of the land, also add to the natural value and become a part of the fixed capital.

Buildings also are looked upon as part of the fixed capital. Strictly speaking, only those buildings which add to the producing power of the farm should be included in the capital invested in the farm. The dwelling house, while commonly added to the investment in the farm, is really intended for the personal use of the farmer and his family. Except in so far as it shelters the help employed on the farm, it can add but little to the returns from it. So far as making a statement of the business of the farm is concerned, it would be better were the farm-house inventoried separately from the other buildings and regarded as a personal expense to the farmer, just as the house of the banker in the city is separated from the business of the bank. All other buildings including silos, corn cribs, granaries, and buildings for sheltering the stock and necessary in conducting the farm business, should be included in the inventoried capital of the farm.

Equipment in the way of teams for work purposes; implements; live stock, such as cows, brood sows, sheep and poultry, that are kept for live stock products, are all a part of the permanent equipment, since they are permanently employed and if sold are replaced by other animals for the same purpose. The circulating or working capital, includes such items of equipment as are frequently changing.

Seed grain, household and farm supplies that are immediately used or marketed, live stock, such as fattening steers, and money for hired labour, are examples of circulating capital. The classificiation intends that the term "circulating capital" shall include only those items that are used once and disappear. If sold for cash, the cash may be invested in other forms of working capital which in turn disappear. Needless to say, the amount of working or circulating capital varies greatly in accordance with the type of business done, with the market, and with the tastes of the farmer. No rule can be given for the exact adjustment of capital for these reasons.

TECHNOLOGICAL NEEDS AND FUTURE AGRICULTURE

It is apparent that the tasks of meeting the consumption needs of the projected population are going to be more difficult given the higher productivity base than in 1960s. There is also a growing realization that previous strategies of generating and promoting technologies have contributed to serious and widespread problems of environmental and natural resource degradation. This implies that in future the technologies that are developed and promoted must result not only in increased productivity level but also ensure that the quality of natural resource base is preserved and enhanced. In short, they lead to sustainable improvements in agricultural production.

Productivity gains during the 'Green Revolution' era were largely confined to relatively well endowed areas. Given the wide range of agroecological setting and producers, Indian agriculture is faced with a great diversity of needs, opportunities and prospects. Future growth needs to be more rapid, more widely distributed and better targeted. Responding to these challenges will call for more efficient and sustainable use of increasingly scarce land water and germplasm resources.

Technical solutions required to solve problems will be increasingly location-specific and matched to the huge agroecological/climatic diversity. Detailed indigenous knowledge and greater skills in blending modern and traditional technologies to enhance productive efficiency will be more than ever before, key to the farming success and sectoral growth. Most technological solutions will have to be generated and adapted locally to make them compatible with socio-economic conditions of farming community.

New technologies are needed to push the yield frontiers further, utilize inputs more efficiently and diversify to more sustainable and higher value cropping patterns. These are all knowledge intensive technologies that require both a strong research and extension system and skilled farmers but also a reinvigorated interface where the emphasis is on mutual exchange of information bringing advantages to all. At the same time potential of less favoured areas must be better exploited to meet the targets of growth and poverty alleviation.

These challenges have profound implications for products of agricultural research. The way they are transferred to the farmers and indeed the way research is organized and conducted. One thing is, however, clear – the new generation of technologies will have to be much more site specific, based on high quality science and a heightened opportunity for end user participation in the identification of targets. These must be not only aimed at increasing farmers' technical knowledge and understanding of science based agriculture but also taking advantage of opportunities for full integration with indigenous knowledge. It will also need to take on the challenges of incorporating the socio-economic context and role of markets.

With the passage of time and accelerated by macro-economic reforms undertaken in recent years, the Institutional arrangements as well as the mode of functions of bodies responsible for providing technical underpinning to agricultural growth are proving increasingly inadequate. Changes are needed urgently to respond to new demands for agricultural technologies from several directions. Increasing pressure to maintain and enhance the integrity of degrading natural resources, changes in demands and opportunities arising from economic liberalization, unprecedented opportunities arising from advances in biotechnology, information revolution and most importantly the need and urgency to reach the poor and disadvantaged who have been by passed by the green revolution technologies.

Another important implication of increasing globalization relates to the need for greater attention to the quality of produce and products both for the domestic and the foreign markets. This would imply that production must be tuned to actual rapidly changing product demand. Such adaptation to global markets would require state of the art research, which can be achieved only by setting global standards of research, focus on well defined priorities and mechanisms which permit close interaction of farmers with researchers, the private sector and markets.

8

Technology of Remote Sensing and Agriculture Resources

Agriculture resources are among the most important renewable, dynamic natural resources. Comprehensive, reliable and timely information on agricultural resources is very much necessary for a country like India whose mainstay of the economy is agriculture. Agriculture survey are presently conducted throughtout the nation in order to gather information and associated statistics on crops, rangeland, livestock and other related agricultural resources. These information of data are most importance for the implementation of effective management decisions at local, panchayat and district levels. In fact, agricultural survey is a backbone of planning and allocation of the limited resources to different sectors of the economy. With increasing population pressure throughout the nation and the concomitant need for increased agricultural production (food and fiber crops as well as livestock) there is a definite need for improved management of the nation agricultural resources. In order to accomplish this, it is first necessary to obtain reliable data on not only the types, but also the quality, quantity and location of these resources. The remote sensing techniques has been and it will continue to, a very important factor in the improvement of the present systems of acquiring and generating agricultural data.

SPACE INPUTS FOR PRECISION AGRICULTURE

Precision agriculture, newly emerging agricultural management concept, embodies the convergence of biotechnologies & other agricultural technologies with space and informatics. With a goal to achieve the quantum jump in agricultural productivity, reduced cost of cultivation, diversified and resilient agricultural systems, the precision agriculture plays catalytic role in order to achieve a common ground between environmental and economic goals. It is basically designed to optimize agricultural inputs viz., fertilizers, pesticides, water etc, in tune with micro-level/field requirements. Optimization is focused on increased yields, reduced cost of cultivation and to minimized environmental impacts through location-specific management.

The success stories pertaining to Precision Agriculture have mainly drawn from the developed countries; wherein Agriculture is characterized by highly mechanized and automated systems, and is driven by market forces and has been professionally managed enterprise. Taking into account the predominance of fragmented land holdings, heterogeneity of crops and livestock and concept of farm families in the rural conditions, the model of Precision Agriculture representing the typical Indian Agricultural scenario is yet to evolve. While the ecological integrity of farming systems is an imperative need, it is equally important to extend the access of information and market to the small and marginalized farmers. The Precision Agriculture model for India while addressing these issues should provides an innovative route for sustainable agriculture in globalize and liberalized economy.

TECHNOLOGY ELEMENTS

Recent advances in technology for variable rate technology (VRT), with concurrent advances in remote sensing, GIS & GPS, and the developments taken place in crop simulation modelling, have provided enough opportunities and scope to take-up proto-type Precision Agriculture experiments. The VRT applies production inputs at rates appropriate to soil and crop conditions within micro-level field conditions. The VRT systems have been demonstrated for several materials, including herbicides, fertilizers, insecticides and seeds.

Role of space technologies becomes more crucial in order to address the spatial variability of soils and crops across the various scales of mapping. The space technology inputs also capture the vulnerability and dynamism of agricultural systems. The developments in space-borne imaging sensors, particularly their spatial, spectral and temporal resolutions are well characterized to capture these features. While high spatial resolution images enable mapping and monitoring the structural attributes of agro-ecosystems, high spectral resolution or hyper-spectral imaging addresses their functionalities. The high temporal resolution captures the dynamisms of agro-ecosystems.

The use of remote sensing, GIS and GPS for crop monitoring, condition assessment and yield modelling has already been well established. Crop simulation models (CSMs) provide potential production under the different scenario of constraints, including weather, soils, crops, cultural practices etc. The conjunctive use of VTR, remote sensing, GIS, GPS and CSMs provides technological framework for Precision Agriculture.

SPACE APPLICATIONS TO AGRICULTURE

Agriculture has been at the top of our priorities for space applications. The road map of agricultural applications started with the first remote sensing experiment on identification of coconut root-wilt diseases in Kerala using Infrared aerial photography-way back in 1969-70. Since then, space applications to agriculture sector have touched almost all the segments of agricultural ecosystems. These include the mapping and monitoring of major crops, soils/ degraded lands, command areas, wastelands, surface & ground water, floods & drought, and watersheds.

Agricultural statistics that provides the vital informatics base to agriculture sector originates from age-old village level Patwari system. It often moves upward with its inherent subjectivity and bias. There are hundreds of crucial decisions at different levels, which are taken purely based on this. In order to strengthen the foundations of agricultural statistics in the country, a remote sensing based the nationwide mission called pre-harvest Crop Acreage and Production Estimation (CAPE) was launched in late 80s.

Covering all the major cereals, pulses and oilseeds, CAPE provides in-season crop statistics with 90/90 accuracy at state level. The CAPE could thus provide the scientific basis of agricultural statistics and is transitioning further to yet another institutional destination entitled Forecasting Agricultural Output using Space Agro-meteorology and Land-based Observations (FASAL) within Ministry of Agriculture itself. Synthesizing the state-of-the-art in econometrics, agricultural meteorology and remote sensing based modelling, FASAL envisages the multiple productions forecasting of the major crops with improved accuracy to the extent of 95/95 criteria, timeliness and scope in terms of covering the whole country. The FASAL could be used to provide scientific solutions facilitating crop insurance, bridging the gaps between crop production and post-harvest technology, pricing and policy decisions.

Extensions of irrigation, genetically improved crops and use of inorganic fertilizers have been the harbingers of Green Revolution in the country. Those multi-purpose irrigation commands, which accelerated the irrigation networks in 70s & 80s, are now characterized by depleted irrigation efficiency, water logging and salinity.

At the behest of Ministry of Water Resources, remote sensing based command areas inventory has been taken up in the priority irrigation commands to enhance the water use efficiency. In the similar line, a coarse scale land degradation mapping was carried out at behest of Ministry of Agriculture. It is important to mention here that in some of critically affected areas of Gangetic plains of Uttar Pradesh, land degradation to the extent of village level has been mapped out. Based on this, several land reclamation efforts-including the World Bank supported programmes, have been taken up to restore the health of soils. In coming years, with the support from Ministry of Agriculture, ISRO/DOS is planning to launch Nationwide Land Degradation Mapping Mission at the scale 1: 50,000, so as to delineate land degradation at Village level.

By virtue of the unique combinations of IRS satellites with varied resolutions and capabilities, the space segment in the country is in a position to monitor agricultural drought and recurrent floods more efficiently. Capturing the flood events with the damages to the Villages and crops has formed the scientific basis for damage assessment and thereby relief operations at various levels. This year, for example, flood inundation maps in Orissa were generated depicting the marooned areas during the different flood waves and disseminated to the user community within hours through the Internet. Space inputs to agricultural drought monitoring is yet another aspect to target the districts and taluks on the basis of severity of drought.

Department of Space at the behest of Ministry of Rural Development has carried out nation-wide wasteland mapping on 1: 50,000 scale using the IRS data. The Wasteland Atlas of India covering the wasteland statistics of entire country has been brought out. These maps would help in retrieving the information at village/watershed (500 ha) levels, for implementation of wastelands/watershed development programmes. The uniqueness of wasteland mapping has also been the creation of digital database with village as well as watershed boundaries.

With the hydro-geomorphological maps prepared under the National Drinking Water Mission, and subsequently to the Rajiv Gandhi National Drinking Water Mission, satellite remote sensing has really made a dent in reaching its potentials down the line to grassroots. Search for groundwater, particularly in areas with consolidated and semi-consolidated rock formations, considered more difficult from the point of view of exploration as well as recharging of groundwater, is considerably aided by the use of the hydro-geomorphological maps. These maps are extensively used for locating prospective groundwater sites around problem habitations in the country, as a part of the 'scientific-source finding'.

At the behest of the Planning Commission, scientific inputs were generated from remote sensing on land use/land cover of entire country to evolve agro-climatic zonation based planning. These applications have moved further to large scale mapping across the problem regions of the country (especially the rain-fed regions), towards micro-level decision making for sustainable development. Under remote sensing based Integrated Mission for Sustainable Development Mission (IMSD), land and water resources development plans were carried out for about 80 mha. in 175 districts of the country. Implementation of these plans in certain areas has enriched the ground water potential, increased the cropping intensity along with the net returns from the fields. It is important to note that these experiments were conducted in the terrain of low productivity, mainly dry lands/rain-fed areas, predominantly wastelands, where the scope of doubling food production is the main issue. Emphasis is being laid in operational utilization of this vast database for ground level implementation by concerned user agencies.

The success of these experiments lies not only due to the applications of space inputs alone, but largely due to the synergistic interfaces among grass root level farmers, administrators/policy makers and multi-disciplinary scientific communities. With operationalisation of the National Natural Resources Management System (NNRMS) in the country, with DOS as the nodal agency, the space, ground and user segments have been properly tuned to respond to the challenges of sustainable crop production in the country.

The INSAT-VHRR observations on clouds, cyclone depressions, surface radiance and monsoon parameters in the Indian Ocean have really strengthened the agro-meteorological services in the country. To disseminate the appropriate locale-specific agricultural packages, there are several programmes such as 'Krishi Darshan' being beamed through all the regional channels. An operational system of public instruction has been established since 1995, using the INSAT based Training and Development Communication Channel (TDCC) for disseminating the improved agricultural practices, training primary school teachers, Panchayat Raj, elected representatives, Anganwadi workers, wasteland development functionaries etc. The Jhabua Development Communication Project (JDCP), undertaken by ISRO in Jhabua district of MP, has been providing communication support to the developmental activities and also interactive training to the development functionaries, besides empowering the poor and marginalised tribal.

A pilot project sponsored by the Planning Commission, a single window information service provider concept-called Agro-climatic Planning and Information Bank (APIB) has been developed for the farmers of drought prone areas of Karnataka. The APIB is an effort to empower the farmers and reduce their vulnerability to crop failures. Moving to the farmer's cooperatives, ISRO has taken up jointly with IFFCO-a major initiative on developing remote sensing and GIS based decision support system for fertilizer movements in the priority areas. Finally, space applications over last three decades in agriculture have made this sector information rich and have addressed the various issues encompassing the different sectors of agriculture.

Indian Precision Agriculture

Agriculture in India, as we see today, is at the crossroads.

1. On the one hand there are depletions of ecological foundations of the agro-ecosystems, as reflected in terms of increasing land degradation, depletion of water resources and rising trends of floods, drought and crop pests and diseases. On the other hand, there is imperative socio-economic need to have enhanced productivity per units of land, water and time.
2. At present, 3 ha of rain fed areas produce cereal grain equivalent to that produced in 1 ha. of irrigated. Out of 142 million ha. Net sown areas, 92 million ha. are under rain-fed agriculture in the county.
3. From equity point of view, even the record agricultural production of more than 200 Mt is unable to address food security issue. A close to 60 Mt food grains in the storehouses of Food Corporation of India (FCI) is beyond the affordability and access to the poor and marginalized in many pockets of the country.
4. Globally, there are challenges arising from the Globalization especially the impact of WTO regime on small and marginalized farmers.
5. Some other unforeseen challenges could be anticipated global warming scenario and its possible impact on diverse agro-ecosystems in terms of alterations in traditional crop belts, micro-level perturbations in hydrologic cycle and more uncertain crop-weather interactions etc.

At this stage, agriculture needs new paradigms to deal with the present situation. The strategy lies in integration of the dynamic information and scientific knowledge into the management of agro-ecosystems, and thereby optimizing the radiation, water and nutrient usages. Agriculture has to transition from high inputs material inputs to the optimum level, through the appropriate use of information, knowledge and strategies for efficient resource usages. In such case, productivity of agricultural may not be the function of the quantum of agricultural input use alone, but will include information, knowledge and efficiency while managing the agricultural practices.

With the fragmentations of land holdings and predominance of small and marginal farmers, our agricultural systems are basically dis-aggregated farm families. Fundamentally, Precision Agriculture aims at a dis-aggregated micro-level farm management strategy with intense information inputs addressing the variations of soils, crops, water, chemicals, market access etc.

Taking into the present state of Agriculture in the country, Precision Agriculture is absolutely essential in order to address poverty alleviation and food security to a very large cross-section of the population. For example, at present, the average Orissa farmer produces slightly more than one ton of rice per hectare, and keeps little below one tone for his family. But if he can produce four tons, then he has three tons to sell, and more cash in hand. The smaller the farm, the greater the need for a marketable surplus.

Defining Precision Agriculture in Indian Context Taking into account the typical characteristics of Indian Agriculture, the definition of Precision Agriculture, therefore, must encompass the strategy and framework to achieve higher productivity, reduced cost of cultivation by optimization of inputs, and diversified & resilient agricultural systems. All these goals are to be achieved within the typical constraints of India's agro-ecosystems. While the depletion of ecological foundations of farming systems needs to be arrested, the access of information, credit, agricultural inputs and market to the small and marginal farmers are equally crucial. The Precision Agriculture model for must be derive encompassing all these issues, within a broad framework of addressing the negatives of globalization and achieving sustainable agriculture in the long run.

The definition of Precision Agriculture has to be micro-level contextual and should capture the local variabilities, vulnerabilities and dynamisms of agro-ecosystems. For example, Agriculture in Punjab, Haryana & Western UP is characterized by higher productivity (about 4 t/ha), higher use of inputs (irrigation ~ 96%; fertilizer consumption 0.158 t/ha), higher cropping intensities and predominance of medium and large farmers.

Sustainability of such agro-ecosystems has been a cause of concern in the recent times. The model for Precision Agriculture for such ecosystems may focus on sustainability through the optimization of agricultural inputs and thereby reduction of cost of cultivation, and cropping system analysis.

Agriculture in Southern part of the country-especially Andhra and Tamilnadu is moderate yield (around 2 t/ha), inputs (irrigation ~ 55%; fertilizer consumption 0.12 t/ha) and cropping system based, with predominance of small and marginal farmers. Precision Agriculture for such system should aim at enhancing the productivity based on in situ soil & water conservation.

The Eastern India especially Orissa, though high potential agricultural system is unfortunately characterized by low inputs (irrigation ~ 25%, fertilizer use 0.025 t/ha), low yield (~1.2t/ha) and low cropping intensity, with predominance of marginal and small farmers. A model for Precision Agriculture for Orissa could focus more on poverty alleviation and food security through enhanced productivity and low cost of cultivation. The models for Precision Agriculture in Indian context could therefore be the micro-level and contextual-addressing the empirical issues in the diverse agro-ecosystems of the country. The access and outreach of best practices in Precision Agriculture-as demonstrated by contextual models, to the large cross-sections of the farming community would then help in cultivating the new paradigms in age-old agriculture enterprise in the country.

Precision Agriculture models are not complete, unless the parameters related to empowerment of the farmers; especially small and marginal farmers are integrated. In this context, ISRO has also initiated Gramsat project in Orissa. In the line of JDCP, the Gramsat project aims at empowering the people especially the poor and marginalized, by awareness building and access to information and services. Towards this, a network of one-way video and two-way audio Village Information Kiosks is being developed in the selected blocks of Orissa. The same networks are also planned to facilitate e-governance in the region. Precision Agriculture model should present a synergy that could lead to a holistic mission, focused on agricultural development with the backdrop of present issues and challenges.

GENERATION AND TRANSFER OF TECHNOLOGY

A very high priority will be accorded to evolving new location-specific and economically viable improved varieties of agricultural and horticultural crops, livestock species and aquaculture as also conservation and judicious use of germplasm and other bio-diversity resources. The regionalization of agricultural research, based on identified agro-climatic zones, will be accorded high priority. Application of frontier sciences like biotechnology, remote sensing technologies, pre and post-harvest technologies, energy saving technologies, technology for environmental protection through national research system as well as proprietary research will be encouraged. The endeavour will be to build a well organized, efficient and result-oriented agriculture research and education system to introduce technological change in Indian agriculture. Upgradation of agricultural education and its orientation towards uniformity in education standards, women empowerment, user-orientation, vocationalization and promotion of excellence will be the hallmark of the new policy.. The research and extension linkages will be strengthened to improve quality and effectiveness of research and extension system. The extension system will be broad-based and revitalized. Innovative and decentralized institutional changes will be introduced to make the extension system farmer-responsible and farmer-accountable. Role of Krishi Vigyan Kendras (KVKs), Non-Governmental Organizations (NGOs), Farmers Organizations, Cooperatives, corporate sector and para-technicians in agricultural extension will be encouraged for organizing demand-driven production systems. Development of human resources through capacity building and skill upgradation of public extension functionaries and other extension functionaries will be accorded a high priority. The Government will endeavour to move towards a regime of financial sustainability of extension services through effecting in a phased manner, a more realistic cost recovery of extension services and inputs, while simultaneously safeguarding the interests of the poor and the vulnerable groups.

Mainstreaming gender concerns in agriculture will receive particular attention. Appropriate structural, functional and institutional measures will be initiated to empower women and build their capabilities and improve their access to inputs, technology and other farming resources.

Inputs Management

Adequate and timely supply of quality inputs such as seeds, fertilizers, plant protection chemicals, bio-pesticides, agricultural machinery and credit at reasonable rates to farmers will be the endeavour of the Government. Soil testing and quality testing of fertilisers and seeds will be ensured and supply of spurious inputs will be checked. Balanced and optimum use of fertilizers will be promoted together with use of organic manures and bio-fertilizers to optimize the efficiency of nutrient use.

Development, production and distribution of improved varieties of seeds and planting materials and strengthening and expansion of seed and plant certification system with private sector participation will receive a high priority. A National Seed Grid will be established to ensure supply of seeds especially to areas affected by natural calamities. The National Seeds Corporation (NSC) and State Farms Corporation of India (SFCI) will be restructured for efficient utilization of investment and manpower.

Protection to plant varieties through a *sui generis* legislation, will be granted to encourage research and breeding of new varieties particularly in the private sector in line with India's obligations under TRIPS Agreement. The farmers will, however, be allowed their traditional rights to save, use, exchange, share and sell their farm saved seeds except as branded seeds of protected varieties for commercial purpose. The interests of the researchers will also be safeguarded in carrying out research on proprietary varieties to develop new varieties. Integrated pest management and use of biotic agents in order to minimize the indiscriminate and injudicious use of chemical pesticides will be the cardinal principle covering plant protection. Selective and eco-friendly farm mechanization through appropriate technology will be promoted, with special reference to rainfed farming to reduce arduous work and to make agriculture efficient and competitive as also to increase crop productivity.

AGRICULTURAL PLANNING AND DEVELOPMENT

India is a vast country with a variety of landforms, climate, geology, physiography, and vegetation India is endowed with regional diversities for its uneven "economic and agricultural" development, on account of (i) Agro-climatic environments (15 Zones/127 regions), (ii) Agro-ecological regions (20) and 60 sub-regions, (iii) Agro-Edephic regions, (iv) Terrain mapping sub-units, (v) Natural resources endowments (geology, geomorphology, soil, ground water, surface water, & infrastructure), (vi) Human resources (Population density), (vii) Level of investments in rural infrastructure, and (viii) Level of investment in technology and its adoption. India has a total geographical area (TGA) of 329 Million Hectares (MH) out of which, about 265 MH represent varying degrees of potential for biological production. [Dhuruva89] report reveals that more than 50% of TGA is threatened by various types of land degradation, such as soil erosion, gully & ravine formation, salinity, water logging, shifting cultivation, etc. Development of irrigation potential is considered as the key factor in the sustenance of "Green Revolution". Despite 50 years of development planning, rainfed agriculture is the largest and the most important sector of crop production in India.

Soil resources are the most precious non-renewable vital resources for growing food, fibre, and fuel wood to meet the human needs. Management of Soil Resources is essential for both the continued agricultural productivity and protection of environment. By considering various factors like population growth rate, diminishing per capita of land and water resources, and increasing land degradation problems, it is estimated that India will be required to produce an additional 5-6 million tons of food grains annually in 21st Century. This will lead to tremendous pressure on soil resources along with competitive demand for it from industrialization and urbanization. However the capacity of soil to produce is limited and its limits to production are set by its inherent characteristics, agro-ecological settings, and its use and management.

the world to preserve the production potential of agricultural land while protecting the environment (FAO, 1993). An integrated approach for sustainable agricultural development requires a combination of agricultural status, surface and groundwater potential, soils, natural vegetation, terrain characteristics, wetland etc. with meteorological information and socio-economic conditions. According to different experts, within the next decade, conventional field surveys will be phased out due to prohibiting costs and likely to be replaced by tools of Information Technology and remote sensing (Bourrough, 1986) to meet the increased demand for up-to-date and reliable information.

Agriculture has always been India's most important economic sector. Large area yield forecasting prior to harvest is of interest to government agencies, commodity firms and producers. Early information on yield and production helps in such activities as planning the type of crops (wheat, rice, barley, oats, etc.). Timely monitoring during the growing season provides opportunities to mitigate any revealed challenges. Information Technology, Remote Sensing Applications, and Decision Support Systems will facilitate decision makers' and farmers' knowledge of the state and situation of the agricultural system, thus helping them plan and act accordingly, Moni, 2003). Information and Communication Technology is a way to link the farmers to Universities and Research Centers, and researchers can orient their work to solve local problems in agriculture (Punjab Agricultural University, 1998).

Some forms of Agriculture Decision Support Systems have been discussed or implemented in different parts of the world. The Crop Condition Assessment Program (CCAP), (Stan Aronoff 2005) is a GIS based analytical Web application that shows in near real-time crop conditions within the Canadian prairies and the US. The MARS (Monitoring Agriculture with Remote Sensing) project (MARS Stat, 2004) focuses mainly on predicting the production volumes of the major crops at a national level and at a regional level for all EU member states. Their methodology is based on remote sensing, agro-meteorological data and crop growth model. The 'Global Monitoring for Environment and Security' (GMES) program is a European initiative and has a component which aims at supporting and complementing existing regional information and early warning systems on food and agriculture.

The Global Information and Early Warning System on Food and Agriculture operated by the United Nations Food and Agriculture Organization (FAO) monitors famine-prone areas of the globe in order to predict food shortages and possible famine conditions. The Production Estimates and Crop Assessment Division (PECAD) of the US Department of Agriculture (USDA) relies on several different data sources to monitor weather anomalies that affect crop production and quality of agricultural commodities (Tetrault and Baldwin,2006). The main agro-meteorological input data sources are the ground meteorological station measurements, and grid weather data integrated with satellite imagery. In addition, several crop models and data reduction algorithms are using both the station and satellite agro-meteorological data sets. These models include crop calendars, crop hazards, and several different crop yield reduction models to assess crop conditions. In India, the concept of an Agricultural Resource Information System (AgRIS) has been extensively discussed (Moni, 2003). The author expects the AgRIS Project to emerge as the richest "agricultural resources information system" covering all aspects of agriculture, natural resources, and food systems, linking the farmers to Universities and Research Centres, whereby researchers can orient their studies to solve local problems in agriculture. In order to develop an Agricultural GeoCapacity Center for India, the AGCN project has been initiated by an Indo-Canadian partnership. With the participation of the Canadian International Development Agency (CIDA), IES and PCI represent the Canadian team, while Punjab Remote Sensing Center (PRSC) and PAU (Punjab Agricultural University) represent the Indian counterpart. Punjab, the home of the 'Green Revolution' in India, has been selected as a pilot area for the AGCN. After several discussion sessions organised in Ludhiana with PRSC and scientists at PAU, the following list of "burning issues" or main concerns for Punjab agriculture were identified:

1. Declining water table and extensive use of rice-wheat cropping system.
2. Declining soil health and increasing soil-water pollution.
3. Improvement of agro-meteorological weather forecasts. Maximization of rainfall agriculture and crop diversification.
4. Better use of rice-wheat straw and protection of environment.
5. Improvements in soil-water-nutrient management.
6. Integrated insect-pest-disease management.
7. Enhancement in crop productivity, quality of produce and sustainability of productivity.
8. Improvement of socio-economic status of Punjab farmers including issues of marketing, mechanization, land parcel size and others.
9. Land use changes and diversification of cropping patterns/farming systems.
10. Development of post-harvest technologies and value additions through processing.
11. Development/training of scientific human resources.

Based on these requirements, AGCN has been designed as an ensemble of tools and applications to provide information for:

- Optimisation of water use for irrigation:
- Precise weather forecasts
- Maximization of rainfall agriculture and crop diversification
- Better planning of:
- Land use
- Farm input, storage and output
- Better modelling and analysis of the crop system based on a multitude of inputs including weather, soil characteristics, population needs, and other economical market parameters
- Guidelines and strategic decisions based on local geographical, industrial and socio-economical characteristics of the area.The proposed Agricultural GeoCapacity Center Network project entails the establishment of a nation-wide information technology system for collecting, collating, organizing and interpreting information on land and resources collection and managing socio-economic data.

The proposed AGCN system, implements the above recommendations. It has enterprise architecture with core GeoCapacity services and data management. The enterprise architecture meets requirements, such as data distribution for efficient sharing and storage, services-oriented architecture for integrating best-of-breed technologies, multi-tier architecture for efficient and effective system enhancements, and environment componentization for system extensibility and scalability. The core of GeoCapacity services and data management provides fully automatic satellite data processing, centralized spatial data storage and management, and the interoperability required to access and use data and information from the various sources.

Bibliography

Allen J.: *Cultural Industries and the Production of Culture,* Taylor & Francis, 2007.

Arrhenius, Gustaf : *Evolution of the Solar System,* Washington D.C. NASA, 1976.

Bhatta, B : *Remote Sensing and GIS,* Oxford University Press, Delhi, 2008.

Boardman, D.: *Geography with Slow Learners,* Sheffield, The Geographical Association, 1982.

Bolstad, P. : *GIS Fundamentals, A first text on Geographic Information Systems, Second Edition,* White Bear Lake, MN, Eider Press, 2005.

Carole L. : *Toward a Locational Definition of State Systems of Settlement,* USA, 1976.

Clark, J.R.: *Coastal Seas: The Conservation Challenge,* Oxford, UK: Blackwell Science, 1998.

Dicken, P. : *Wheels of Change, the Automobile Industry, Global shift, Mapping the Changing Contours of the World Economy,* London, 2007.

Earney, F.C.F.: *Marine Mineral Resources: Ocean Management and Policy,* London, Routledge, 1990.

Elachi, Charles : *Introduction to the Physics and Techniques of Remote Sensing,* New York, Wiley, 1987.

Gallagher, K.S., : *China Shifts Gears, Automakers, Oil, Pollution and Development,* MIT Press, Cambridge, Mass, 2006.

Gomez, A.A., Swete Kelly: *Measuring Sustainability of Agricultural Systems at the Farm,* Madison, USA, 1996.

Goodchild, M.F and Kemp, K.K: *National Centre for Geographic Information and Anlysis,* Univ. of California, USA, 1990.

Gurney, R. J. : *Remote Sensing in Hydrology,* New York, Van Nostrand Reinhold, 1991.

Harvey, Francis : *A Primer of GIS, Fundamental Geographic and Cartographic Concepts,* The Guilford Press, 2008.

Jamwal, Anil : *Principles of Remote Sensing,* Jnanada Prakashan, Delhi, 2010.

Kiefer, R. W. : *Remote Sensing and Image Interpretation,* New York, 1987.

Kitchin, R.: *Geography and Disability,* Sheffield, Geographical Association, 2000.

Kumar Yadav, Surendra : *Soil Ecology : Remote Sensing and Geographic Information System in Management,* APH, Delhi, 2007.

Mack, Pamela Etter : *Viewing the Earth: The Social Construction of the Landsat Satellite System,* Cambridge, MIT Press, 1900.

Matthews, M. C. : *Remote Sensing in Civil Engineering,* Glasgow, Surrey University Press; New York, 1985.

Maxton, G.P. and Wormald, J. : *Time for a Model Change, Re-engineering the Global Automobile Industry,* Cambridge University Press, New York, 2004.

McDonnell, R.A. : *Principles of Geographical Information Systems*, Oxford University Press, Oxford, 1998.

Nicholls, A.: *Low Attainers and the Teaching of Geography*, Sheffield, Geographical Association, 1988.

Nierenberg, W. A.: *Encyclopedia of Earth System Science*, San Diego, California: Academic Press, 1992.

Okoshi, T. : *Three-dimensional Imaging Techniques*, Academic Press Inc., New York, 1976.

Ramasamy, S. : *Remote Sensing in Water Resources*, Rawat, 2005.

Rawling, E.: *Teaching Slow Learners through Geography*, Sheffield, GeographicalAssociation, 1985.

Raymond M. : *Laser Remote Sensing: Fundamentals and Applications*, Malabar, Fla., Krieger Pub. 1992.

Rondinelli, D.A.: *Applied Methods of Regional Analysis: the Spatial Dimension of Development Policy*, Westview Press, Boulder and London, 1985.

Ruth Towse : *Developments in the Economics of Copyright, Research and Analysis*, Edward Elgar Publishing, Cheltenham, United Kingdom, 2005.

Scott, Allen J.: *Cultural Industries and the Production of Culture*, Taylor & Francis, 2007.

Sebba, J.: *Geography for All*, London, David Fulton, 1996.

Sokhi, B.S. and S.M. Rashid : *Remote Sensing of Urban Environment*, Manak, Delhi, 1999.

Steven J.: *Ecopreneuring: The Complete Guide to Small Business Opportunities from the Environmental Revolution*, John Wiley and Sons, 1992.

Steven, A. : *Grand Canyon: Monument to Catastrophe*, Institute for Creation Research, London,1994.

Thomashow, Mitchell: *Ecological Identity: Becoming a Reflective Environmentalist*, Cambridge, MIT Press, 1995.

Vogel, Harold L. : *Entertainment Industry Economics, A Guide for Financial Analysis*, Cambridge University Press, 2004.

Walter Leal: *Innovation, Education, and Communication for Sustainable Development*, New York, Peter Lang Publishing, 2006.

Wendy J. Gordon : *The Economics of Copyright, Developments in Research and Analysis* Edward Elgar, 2003.

Index

S

T

❑❑❑